Classical Theory of Free-Electron Lasers

A text for students and researchers

Classical Theory of Free-Electron Lasers

A text for students and researchers

Eric B Szarmes

Department of Physics and Astronomy, University of Hawai'i at Mānoa, Honolulu, HI, USA

Morgan & Claypool Publishers

ISBN 978-1-6270-5573-4 (ebook)
ISBN 978-1-6270-5572-7 (print)
ISBN 978-1-6270-5680-9 (mobi)

DOI 10.1088/978-1-6270-5573-4

Version: 20141201

IOP Concise Physics
ISSN 2053-2571 (online)
ISSN 2054-7307 (print)

A Morgan & Claypool publication as part of IOP Concise Physics
Published by Morgan & Claypool Publishers, 40 Oak Drive, San Rafael, CA, 94903, USA

IOP Publishing, Temple Circus, Temple Way, Bristol BS1 6HG, UK

Dedicated in loving memory of my father, Kornel Rudolf Victor Szarmes

Contents

Preface x

Acknowledgements xii

Author biography xiii

1 Introduction and overview 1-1

1.1 The free-electron laser 1-1
1.2 Classical stimulated emission 1-3
1.3 Electron bunching 1-6
1.4 FEL equations of motion 1-7
References 1-8

2 The classical limit 2-1

2.1 Emission and absorption 2-1
2.2 Compton recoil 2-4
2.3 Wavepacket spreading 2-4
References 2-5

3 Electron beam dynamics 3-1

3.1 Phase space and emittance 3-1
3.1.1 Beam envelope equation 3-3
3.2 Focusing properties of the undulator 3-4
3.3 Matching into the FEL 3-6
Reference 3-9

4 Undulator trajectories 4-1

4.1 Transverse motion 4-1
4.2 Longitudinal motion 4-2

5 Spontaneous emission 5-1

5.1 Spectral lineshape 5-1
5.2 Spontaneous power (weak undulator fields) 5-2
5.3 Spontaneous power (strong undulator fields) 5-5
References 5-6

6	**Effect of the optical field on electron motion**	**6-1**
6.1	The Lorentz equation	6-1
6.2	The FEL pendulum equation	6-3
	References	6-4
7	**Effect of electron motion on the optical field**	**7-1**
7.1	The wave equation	7-1
7.2	Transverse currents	7-4
7.3	The FEL wave equation	7-5
7.4	Energy conservation	7-7
	References	7-8
8	**Transverse modes in the equations of motion**	**8-1**
8.1	Superposition of transverse modes	8-1
8.2	The mode evolution equation	8-3
8.3	The multimode pendulum equation	8-4
8.4	The filling factor	8-6
	References	8-9
9	**Small-signal gain—first derivation**	**9-1**
9.1	Gain from energy conservation	9-1
9.2	Gain-spread theorem	9-7
9.3	Approximate solution of the FEL equations	9-8
9.4	Gouy phase shift	9-12
	References	9-15
10	**Gain reduction and other effects**	**10-1**
10.1	Electron beam emittance	10-1
10.2	High current and high gain	10-3
10.3	Energy spread	10-6
10.4	Short-pulse effects	10-8
10.5	Summary	10-9
	Reference	10-11
11	**Laser saturation and output power**	**11-1**
11.1	The nature of FEL saturation	11-1
11.2	Strong-saturation effects	11-2

11.3 Intensity dependence 11-5
11.4 Analysis of optical resonators 11-7
11.5 Extraction efficiency 11-10
11.6 Incorporation of energy spread 11-13

12 Harmonic lasing 12-1

12.1 Small-signal gain 12-1
12.2 Saturation and output power 12-4
12.3 Spontaneous emission 12-6

13 Helical undulators 13-1

13.1 Electron trajectories 13-1
13.2 FEL coupled equations of motion 13-3
13.3 Small-signal gain 13-5

14 Small-signal gain—second derivation 14-1

14.1 The equation for weak fields 14-1
14.2 FEL gain and dispersion 14-4
14.3 A digression on numerical simulations 14-5
References 14-6

15 Short-pulse propagation 15-1

15.1 General description 15-1
15.2 The coupled Maxwell–Lorentz equations 15-2
15.3 Optical pulse evolution 15-4
15.4 Cavity detuning and refractive effects 15-6
15.5 Mode locked FEL theory 15-7
References 15-16

Preface

This textbook grew out of a set of handwritten notes that I originally wrote and compiled as an instructor for Physics 245, Free-Electron Lasers (FELs), at the US Particle Accelerator School in 1996, and have presented as a component of Physics 660, Advanced Optics, at the University of Hawai'i for over ten years. The text focuses on the fully classical theory of FELs with application to FEL oscillators and develops the fundamentals of FEL theory in sufficient depth to provide both a solid understanding of FEL physics and a solid background for research in the field. The topics have evolved over the years and have been reorganized and augmented with new content since their original presentation. Revisions include a correction in the calculation of the small-signal gain, an extended analysis of saturation and new sections on the classical limit, electron beam dynamics, harmonic lasing and helical undulators. All numerical approximations were developed by the author and numerous examples are included throughout to illustrate the application of analytical results. In conformity with most of the early literature on FELs, centimeter-gram-second (CGS) units are employed for all equations and physical quantities except where explicitly noted.

The text is written at a level suitable for advanced undergraduate or graduate students. Students should have taken a course in advanced electrodynamics at the level of Zangwill or Jackson (1975), including exposure to the theory of physical optics and Gaussian beams, and be familiar with relativistic mechanics and electrodynamics. Students should also have a good working knowledge of the techniques of higher mathematics including partial differential equations, Fourier theory, orthogonal functions and complex analysis.

The pedagogic aim of this work is to provide a coherent description of classical FEL physics of practical utility for students and researchers; it is not intended to serve as a review of the literature. For this purpose the references by Brau (1990), Colson (1990) and Friedman *et al* (1988) provide comprehensive overviews of the field and contain extensive references to original research. Specific citations for the current text include Siegman (1986) for general laser and resonator theory, Jackson (1975) for electrodynamics and selected publications from the literature on FEL theory. The analysis of saturation in chapters 11 and 12 and the theory of mode locking in chapter 15 are, to the best of my knowledge, original contributions of the author. Although I cannot claim priority for the other calculations presented in the text, which date from the earliest days of FELs, all results were rederived by the author with special attention to pedagogy and I believe that a number of the derivations are unique. Most of all I hope that the pedagogic organization of the work will be especially helpful to both students and researchers.

All illustrations were created by the author using Canvas and KaleidaGraph for the Macintosh. Special care was taken in the artwork to employ actual mathematical shapes for all relevant functions and curves.

Honolulu, HI E B Szarmes
September 2014

References

Brau C A 1990 *Free-Electron Lasers* (Boston, MA: Academic)

Colson W B 1990 Classical free electron laser theory *Laser Handbook* vol 6, ed W B Colson *et al* (Amsterdam: North-Holland) chapter 5

Friedman A, Gover A, Kurizki G, Ruschin S and Yariv 1988 A spontaneous and stimulated emission from quasifree electrons *Rev. Mod. Phys.* **60** 471–535

Jackson J D 1975 *Classical Electrodynamics* 2nd edn (New York: Wiley)

Siegman A E 1986 *Lasers* (Mill Valley, CA: University Science Books)

Zangwill A 2013 *Modern Electrodynamics* (Cambridge: Cambridge University Press)

Acknowledgements

I am grateful to Mel Month for the opportunity to teach a course on free-electron lasers at the US Particle Accelerator School in 1996, when I first compiled the notes on which this text is based. I wish to express my gratitude to Steve Benson, who taught me FEL theory in the early years, and whose tutelage I look back on with great fondness. I wish to acknowledge helpful and stimulating discussions on fundamental electrodynamics with John Madey and Charles Brau, which significantly sharpened my understanding of the subject. I am especially grateful to John Madey for his invention of the free-electron laser, for his many pioneering contributions to the field, and for the personal opportunity to enjoy a long and fruitful collaboration with him over the years. I wish to express my sincere thanks to my students at the University of Hawaii for their many insightful questions, and to the publishers at Morgan & Claypool for their interest in publishing this text. Most profoundly, I am grateful to my parents for their support and encouragement, and to my father in particular for instilling in me a love of science.

Author biography

Eric B Szarmes

Originally from British Columbia, Canada, Eric Szarmes received his Bachelor of Applied Science in Engineering Physics from the University of British Columbia in 1985, and his PhD in Applied Physics from Stanford University in 1992, where he did his doctoral research in high resolution free-electron laser spectroscopy under Professor John Madey. He was a postdoctoral research scientist at the Duke Free-Electron Laser Laboratory from 1992 to 1998, where he made pioneering contributions to the phase-locked and chirped-pulse free-electron laser. In 1998 he joined the faculty of the University of Hawaii where he is currently an associate professor of physics. His current research interests include the theory and design of novel optical resonators for high-resolution free-electron laser spectroscopy, x-ray generation and high-field physics. His greatest passion is for teaching.

Chapter 1

Introduction and overview

1.1 The free-electron laser

A free-electron laser (FEL) is a laser source that produces spatially and temporally coherent optical radiation by stimulated emission, where in place of an atomic or molecular medium to provide amplification the gain medium is comprised of a beam of relativistic electrons traveling in a vacuum through a periodic magnetic field. The basic components common to all FELs are a relativistic electron beam, a periodic magnetic structure (an undulator or wiggler magnet of spatial period λ_w), and an optical resonator providing feedback and amplification. (X-ray FELs such as the Linac Coherent Light Source at Stanford omit the optical resonator by necessity and achieve the required gain on a single pass.) The features that make FELs particularly useful as research devices are the unique combination of continuous and broadband tunability, high peak and average power, and spatial and temporal coherence.

A wide range of FEL technologies currently exist, the specific choice of which is determined primarily by the desired operating wavelength—centered anywhere between the x-ray and far-IR regimes—and the associated accelerator technology. This text is developed in the context of the MkIII RF-linac FEL illustrated in figure 1.1, which produces continuously tunable laser light within the IR band delimited by the Nd:YAG laser at 1.06 μm and the CO_2 laser at 10.6 μm. The configuration of this system, apart from the physics of the FEL interaction itself, also impresses several technology-dependent properties on the laser light that is produced.

The FEL optical wavelength λ depends on the electron energy γmc^2 and the undulator field strength $a_w \propto B$ by the tuning relation $\lambda = \frac{\lambda_w}{2\gamma^2}[1 + a_w^2]$; continuous tunability is afforded by continuous variation of either of these parameters. In practice, the electron beam optics used to transport and focus the electron beam into the laser are also energy dependent and tuning is typically achieved at fixed γ by adjusting the undulator magnetic field B. The parameter a_w^2 in the MkIII FEL provides useful gain for values between ~0.3 and 1.4, thus allowing almost a full

doi:10.1088/978-1-6270-5573-4ch1

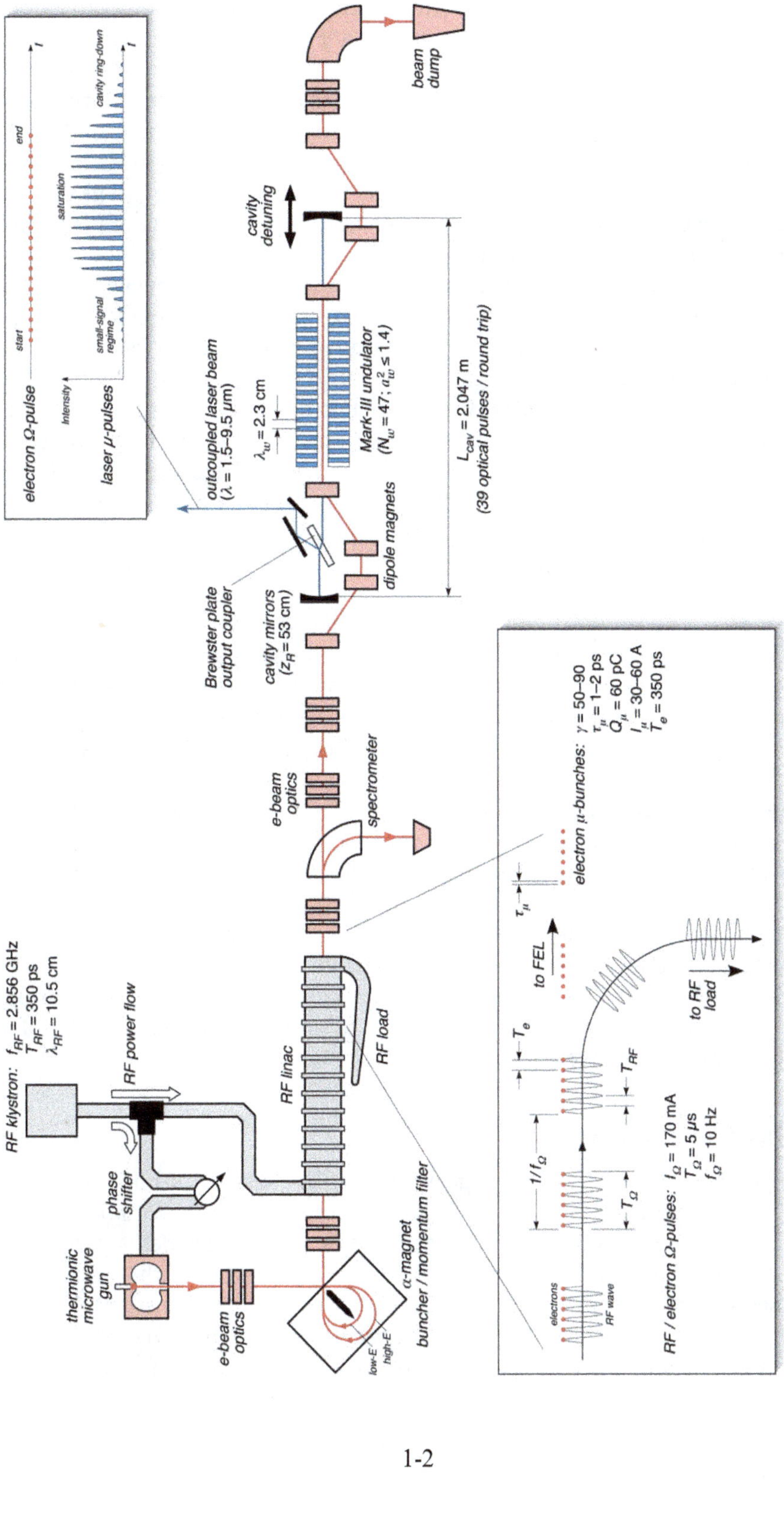

Figure 1.1. The MkIII RF-linac FEL.

octave of tuning at fixed energy. To accommodate broadband tunability, the optical cavity employs metal mirrors with Brewster plate output coupling of the linearly polarized light.

Since the electron beam acts as the gain medium in an FEL, the pulse format of the co-propagating optical beam largely replicates the electron pulse format. In the MkIII FEL, the accelerator is an RF linear accelerator (RF-linac) fed by a 10 Hz, pulsed RF klystron with an RF frequency of 2.856 GHz and a pulse duration of ~5 μs that feeds both the microwave electron gun and the RF-linac. Due to thermionic emission in the gun, electrons are emitted in 'bunches' at the RF frequency. These mildly relativistic electron bunches are filtered and compressed in an *alpha-magnet* buncher to picosecond duration and are subsequently accelerated by the RF-linac to a kinetic energy between 25 MeV and 45 MeV. The electron beam which is transported to the laser thus possesses a dual pulse structure: *macropulses* (Ω-pulses) of ~5 μs duration are cycled at a repetition rate of 10 Hz, with each macropulse consisting of a GHz-rate pulse train of picosecond electron *micropulses* (μ-pulses or μ-bunches) separated in time by 350 ps. This dual pulse structure is ultimately imparted to the laser beam.

When the picosecond electron bunches enter the undulator, they generate and amplify a series co-propagating picosecond optical pulses. To accommodate an undulator long enough to provide useful gain ($N_w = 47$ periods with $\lambda_w = 2.3$ cm), the optical resonator of the MkIII FEL is 2 m long and contains 39 circulating optical pulses. The physical coupling between the electron bunches and optical pulses can thus be described as a type of synchronously pumped harmonic mode locking, in which gain-modulation of the electron beam replaces optical loss- or phase-modulation. Optimum mode locking is achieved by precisely tuning the optical cavity length so that the axial mode spacing equals the 39th subharmonic of the 2.856 GHz modulation frequency; in the time domain, this translates into the requirement that each circulating optical pulse temporally coincide with an injected electron bunch on every round trip. Given the optical round trip time of 13.7 ns, each 5 μs electron Ω-pulse can accommodate almost 400 round trips of laser gain; laser saturation is typically reached within the first microsecond.

In the text that follows, the physics of the FEL interaction is initially developed in the context of continuous wave (CW) electron and optical beams. The theory of pulsed FEL operation and mode locking will be developed through a straightforward generalization of the CW equations of motion. Various aspects of the technology described above are referenced throughout the narrative.

1.2 Classical stimulated emission

Any physical system that can undergo spontaneous emission of photons can also undergo stimulated emission. This fact derives from the general form of the interaction Hamiltonian in quantum electrodynamics, which includes a term in the photon creation operator that formally acts identically on both the vacuum state (spontaneous emission) and the multiphoton state (stimulated emission).

Now, electrons in a relativistic beam propagating through a synchrotron undulator magnet certainly undergo spontaneous emission—synchrotron radiation—and so should also be capable of stimulated emission. The development of this idea lead to the concept and realization of the FEL (Madey 1971).

The wavelength of spontaneous radiation in an undulator is readily derived using a Doppler upshift analysis. Let the undulator be plane-polarized with period λ_w. In the electron rest frame (ERF), assumed to move at constant velocity, the undulator magnetic field is transformed into a linearly polarized incident EM wave with a Lorentz-contracted wavelength

$$\lambda' = \frac{\lambda_w}{\gamma_z}, \tag{1.1}$$

where γ_z is the γ-factor of the ERF in the z-direction, parallel to the undulator axis. This incident EM wave produces transverse dipolar oscillations in the electrons and they radiate the characteristic dipole pattern (Thomson scattering) with a frequency in the ERF given by

$$\omega' = \frac{2\pi c}{\lambda'} = \frac{2\pi c}{\lambda_w}\gamma_z \equiv \gamma_z k_w c, \tag{1.2}$$

where $k_w \equiv 2\pi/\lambda_w$ is the *undulator wavenumber*. In the laboratory frame, this Thomson-scattered radiation is Doppler shifted by another factor of $2\gamma_z$ in the forward direction, yielding spontaneous radiation with a frequency and wavelength given by

$$\omega = 2\gamma_z\omega' = 2\gamma_z^2 k_w c, \qquad k = \frac{\omega}{c} = 2\gamma_z^2 k_w, \qquad \lambda = \frac{\lambda_w}{2\gamma_z^2}. \tag{1.3}$$

Classically, how does spontaneous radiation support the subsequent process of stimulated emission? Observe that we can write

$$\frac{1}{\gamma_z^2} = 1 - \beta_z^2 = (1 + \beta_z)(1 - \beta_z) \simeq 2(1 - \beta_z), \tag{1.4}$$

where $\beta_z \equiv v_z/c$ is the velocity of the ERF in the z-direction and the last approximation is valid for $\gamma_z \gg 1$, $\beta_z \simeq 1$. Equation (1.3) can thus be written

$$\lambda = \lambda_w(1 - \beta_z). \tag{1.5}$$

Let us now examine the relative speeds of the radiation and electron in the laboratory frame. Assume that the electron is initially coincident with a given wavefront of the EM wave. In the time Δt_w that it takes the electron to traverse one magnet period, the radiation wavefront has traveled ahead of the electron by a distance

$$(c - v_z)\Delta t_w = (c - v_z)\frac{\lambda_w}{v_z} = \left(\frac{1}{\beta_z} - 1\right)\lambda_w \simeq (1 - \beta_z)\lambda_w. \tag{1.6}$$

By (1.5), this distance is equal to the wavelength λ of the spontaneous radiation in the laboratory frame. Therefore, after the electron has traversed one magnet period, the

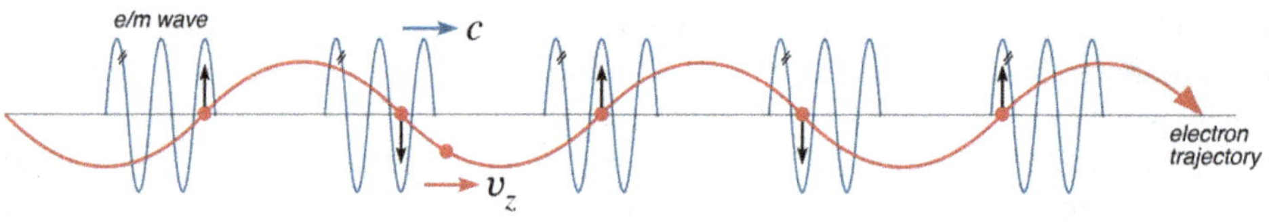

Figure 1.2. The FEL slippage condition.

radiation has slipped past the electron by one optical wavelength. This is the so-called *slippage condition* and motivates the picture of stimulated emission shown in figure 1.2.

We see that the electron's transverse velocity changes sign in each half period of the magnet but, because of optical slippage, so does the $\vec{E}$-vector in the radiation field that interacts with the electron. Thus, for a given electron, the sign of the transverse velocity remains essentially fixed with respect to the sign of the optical $\vec{E}$-field throughout the entire undulator and each electron is poised to continuously lose energy to the field or gain energy from it, depending on the microscopic longitudinal position of the electron within the optical wave. More precisely, the electron energy E at position z evolves as

$$\begin{aligned}\frac{\mathrm{dE}}{\mathrm{d}t} &= \vec{v}\cdot\vec{F} \sim \cos(k_w z)\cos(kz-\omega t)\\ &\sim \cos\left[(k+k_w)z-\omega t\right]+\cos\left[(k-k_w)z-\omega t\right].\end{aligned} \tag{1.7}$$

The second term in this expression travels with a phase velocity greater than the speed of light and thus is nonresonant with the electron; it oscillates at the position of the electron and contributes an average energy change of zero. The first term, however, travels with a phase velocity of

$$v_{\mathrm{pp}} = \frac{\omega}{k+k_w} = \frac{c}{1+\dfrac{k_w}{k}} = \frac{c}{1+\dfrac{\lambda}{\lambda_w}} \simeq c\left[1-\frac{\lambda}{\lambda_w}\right] = c\left[1-(1-\beta_z)\right] = v_z, \tag{1.8}$$

equal to the electron velocity in the z-direction. Consequently, this term remains resonant with the electron along the entire undulator (to zeroth order in the energy change $\Delta\gamma$) and yields a monotonic exchange of energy with the radiation field. We define

$$\xi = (k+k_w)\,z(t) - \omega t \tag{1.9}$$

as the *electron phase in the ponderomotive potential*, where $z(t)$ is the position of the electron. The specific value of ξ determines whether a given electron gains energy from the field or loses energy to the field. The concept of the ponderomotive potential U_{pp} comes from associating a *ponderomotive force* $F_{\mathrm{pp}} \equiv \mathrm{dE}/\mathrm{d}z$ with the resonant term in (1.7),

$$\frac{\mathrm{dE}}{\mathrm{d}t} = v_z\frac{\mathrm{dE}}{\mathrm{d}z} = v_z F_{\mathrm{pp}} = K\cos\left[(k+k_w)z-\omega t\right], \tag{1.10}$$

where F_{pp} acts in the longitudinal direction, and

$$U_{pp} = -\int^{z} F_{pp}(z')\mathrm{d}z' = -\frac{K}{\omega}\sin\left[(k + k_w)z - \omega t\right], \tag{1.11}$$

where we substituted $\omega = v_z(k + k_w)$ from (1.8).

1.3 Electron bunching

Consequently, notwithstanding the resonant condition expressed by (1.8), the phase ξ does evolve slightly, because if electrons are poised to gain energy, their forward speed will increase and they will begin to drift ahead. Electrons poised to lose energy will begin to drift back. These drifts occur on higher orders of the energy change $\Delta\gamma$ and represent motion towards the stable points in the ponderomotive potential.

The electrons therefore become bunched on the scale of an optical wavelength. To illustrate this evolution in electron phase, let us follow a given sample of electrons through the undulator (figure 1.3).

The effect of the resonant interaction is to induce a longitudinal velocity modulation on the scale of the optical wavelength and this modulation correspondingly induces bunching on the same scale by the end of the undulator. The electron bunching mechanism leads to coherent emission and ultimately to amplification and coherence (i.e. well defined phase) in the co-propagating and recirculating optical wave.

Thus, classical stimulated emission ultimately results from, and depends on, the resonant interaction of the wave with the transverse oscillation of the charges and can be described in complete analogy by an EM wave interacting with the oscillation of a charged particle on a spring. This picture, of course, is the same one employed by Siegman (1986) to discuss in classical terms the stimulated emission by coherently oscillating atomic dipole moments in conventional lasers.

Still, we have not quite reached the end of the story. In the above description of the FEL interaction, the radiation field induced bunching on the scale of an optical wavelength, but the bunches coalesced around the stable points in the ponderomotive potential. In this circumstance, the same number of electrons gained energy as lost energy and there was no net energy transfer to the optical wave. Indeed, when

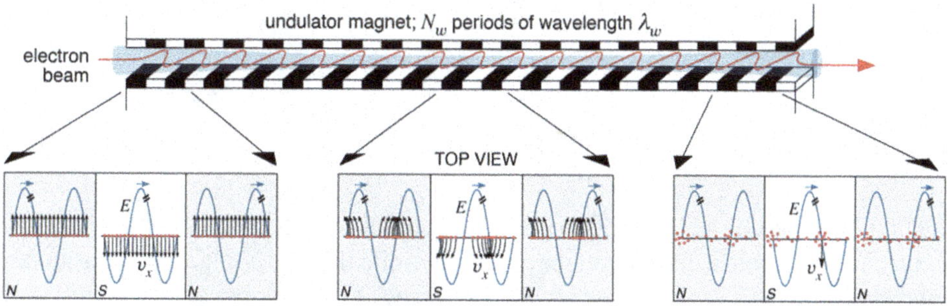

Figure 1.3. Conceptual illustration of the bunching mechanism in an FEL.

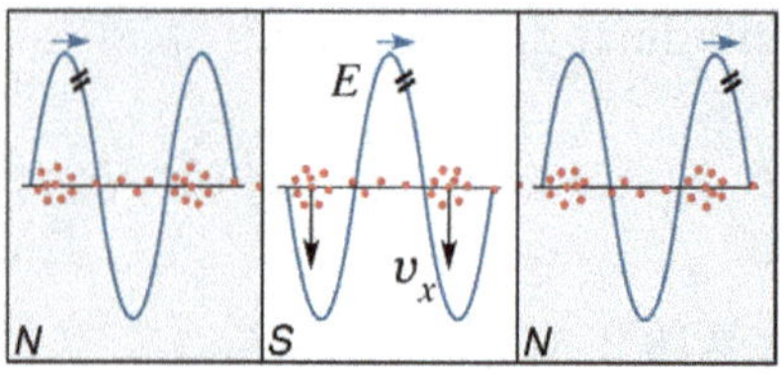

Figure 1.4. Conceptual illustration of the off-resonant gain mechanism in an FEL.

we develop the full FEL theory we will see that there is no gain for an exactly resonant interaction, at least for low currents.

To achieve net gain in the optical wave, we merely have to slightly bias the initial forward velocity of the electrons, so that by the time the electrons become bunched at the end of the undulator this velocity bias will push the electron bunches ahead of the stable points to a non-stationary part of the potential. The result is that electrons whose initial energy is slightly higher than the resonant energy will on average lose energy to the optical wave and the optical wave will be amplified. The effect of this velocity bias is illustrated in figure 1.4; we see that the bunches do actually now drive the wave and increase its amplitude, because they are poised to lose energy on average.

The velocity bias is specified in terms of the time derivative of the phase ξ:

$$\text{phase velocity} = \nu \equiv \frac{\mathrm{d}\xi}{\mathrm{d}\tau} = L_w\left[(k + k_w)\beta_z - k\right], \tag{1.12}$$

where the dimensionless time $\tau = ct/L_w$ varies from 0 to 1 along the undulator of length $L_w = N_w\lambda_w$.

Note that the phase velocity $\nu = \nu(\beta_z(\gamma), k(\omega))$ is a function of the independent variables γ and ω. In an FEL oscillator starting from spontaneous radiation, the injected electron energy is typically held fixed, but the spontaneous spectrum contains a frequency component, offset from the resonant frequency, for which the phase velocity is appropriately biased. It is *that* frequency component which is amplified.

1.4 FEL equations of motion

Ultimately, the fundamental FEL interaction is described in its entirety by two basic equations, the *coupled Maxwell–Lorentz* equations of motion:

$$\frac{\mathrm{d}\nu}{\mathrm{d}\tau} = |a|\cos(\xi + \phi) \tag{1.13}$$

$$\frac{\mathrm{d}a}{\mathrm{d}\tau} = -j\langle \mathrm{e}^{-\mathrm{i}\xi}\rangle_{\xi_0,\nu_0}. \tag{1.14}$$

The first equation is the *FEL pendulum equation*, which describes how electrons evolve in (ξ, ν)-space in the presence of an optical field $a = |a|\mathrm{e}^{\mathrm{i}\phi}$. The second equation is the *FEL wave equation*, which describes how the optical field is driven by the resulting electron distribution in phase space. The course of study we develop in this text will consist of deriving these equations, gaining insight into their meaning and exploring their application to the physics of FELs.

References

Madey J M J 1971 Stimulated emission of bremsstrahlung in a periodic magnetic field *J. Appl. Phys.* **42** 1906–13

Siegman A E 1986 *Lasers* (Mill Valley, CA: University Science Books)

IOP Concise Physics

Classical Theory of Free-Electron Lasers

A text for students and researchers

Eric B Szarmes

Chapter 2

The classical limit

2.1 Emission and absorption

This chapter establishes the basis for the fully classical description of FELs in which both the optical wave and electrons are treated classically. The discussion in this chapter is adapted from the paper by Friedman *et al* (1988). We first consider the kinematics of photon emission and absorption by an electron in an undulator and we assume a weak undulator field, with $\gamma_z \simeq \gamma$, for which the relation between electron energy and momentum is closely approximated by the relativistic equation $E^2 = c^2p^2 + m^2c^4$. Figure 2.1 illustrates the associated kinematics assuming propagation of both the electrons and photons in the $+z$-direction.

We first observe that photon emission and absorption by an electron in free space is forbidden: the respective E–p curves intersect only at the initial state i, so the transition to any final state that conserves energy would violate momentum conservation. (Since the E–p surfaces for the photon and electron are, respectively,

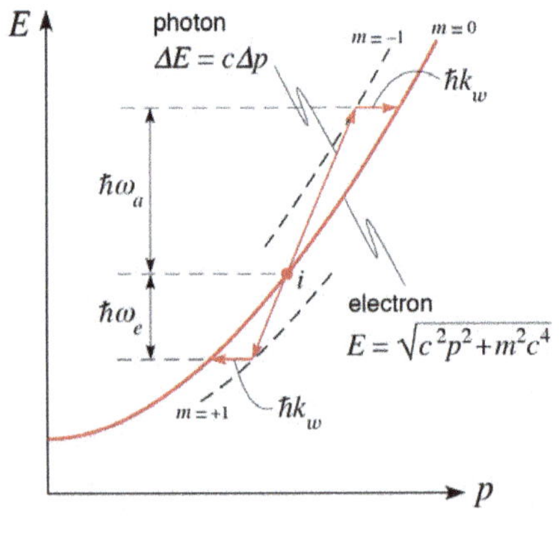

Figure 2.1. Kinematics of photon emission and absorption in an FEL.

doi:10.1088/978-1-6270-5573-4ch2

cones and hyperboloids of revolution, the momentum deficit is even greater for off-axis emission or absorption.)

However, in an undulator magnet, the excess momentum can be absorbed or supplied by the undulator field. Physically, the electron wavefunction assumes the periodicity of the magnetic field, $\psi = \sum_m \psi_m \exp[\mathrm{i}(k_z + mk_w)z]$, where $k_w = 2\pi/\lambda_w$ is the undulator wavenumber. The components of this wavefunction are separated in momentum by integral multiples of $\hbar k_w$, which represent quanta of the periodic magnetic field. Consequently, transitions to states of higher or lower electron energy can occur via states displaced by $\Delta m = \pm 1$ (or $\Delta m = \pm f$ for harmonic radiation) with the momentum deficit supplied by the undulator magnet, as illustrated in figure 2.1. From this figure we immediately see that the fundamental frequencies ω_a and ω_e for absorption and emission satisfy $\omega_e < \omega_a$ in all cases. Quantitatively, we have (with the upper sign corresponding to ω_a)

$$E_\pm^2 = c^2 p_\pm^2 + m^2c^4 \tag{2.1}$$

$$\left(E_i \pm \hbar\omega_{a,e}\right)^2 = c^2\left(p_i \pm \frac{\hbar\omega_{a,e}}{c} \pm \hbar k_w\right)^2 + m^2c^4 \tag{2.2}$$

$$\omega_{a,e}(E_i - p_i c \mp \hbar k_w c) = p_i k_w c^2, \tag{2.3}$$

where we canceled the terms $E_i^2 = c^2p_i^2 + m^2c^4$ and $\hbar^2\omega_{a,e}^2$ from both sides, dropped the term $(\hbar k_w)^2c^2$ with respect to $2p_i(\hbar k_w)c^2$, multiplied the equation for ω_e by -1, and divided both equations by $2\hbar$. If we now substitute $E_i = \gamma mc^2$ and $p_i = \beta\gamma mc$, together with the approximations $1 - \beta \simeq 1/2\gamma^2$ and $\beta \simeq 1$ for relativistic motion, we obtain

$$\omega_{a,e}\left(1 \mp \frac{2\hbar k_w \gamma}{mc}\right) = 2\gamma^2 k_w c. \tag{2.4}$$

In the classical limit as $\hbar \to 0$, we see that the frequencies for emission and absorption converge to

$$\omega_a = \omega_e \equiv \omega = 2\gamma^2 k_w c, \tag{2.5}$$

as we previously established in (1.3). However, even in the classical regime established below, the frequencies ω_a and ω_e remain separated by the tiny fractional shift

$$\frac{\omega_a - \omega_e}{\omega} = \frac{4\hbar k_w \gamma}{mc} = \frac{2\hbar\omega}{\gamma mc^2} \tag{2.6}$$

equal to twice the photon energy divided by the electron energy, where we substituted for k_w in terms of ω from (2.5) in the second equality. The *classical regime* is defined by the relative magnitude of this shift compared to the fractional width of the optical spectrum, which arises from the finite length $L_w = N_w\lambda_w$ of the undulator: the wavenumber k_w in (2.4) actually represents a component of a distribution of wavenumbers $f(k')$, centered on k_w, of shape and width determined by the spatial Fourier transform $\mathcal{F}$ of the undulator field,

$$f(k') = \mathcal{F}\left\{\mathrm{rect}\left(\frac{z}{L_w}\right)\mathrm{e}^{ik_w z}\right\} = L_w\ \mathrm{sinc}\left(L_w\frac{k' - k_w}{2\pi}\right), \tag{2.7}$$

where $\mathrm{rect}(x) = \{1; 0\}$ for $\left\{|x| \leqslant \frac{1}{2};\ |x| > \frac{1}{2}\right\}$ and $\mathrm{sinc}(x) \equiv [\sin \pi x]/[\pi x]$. It is clear from the linear dependence of $\omega_{a,e}$ on k_w in (2.5) that the same distribution governs the photon emission and absorption spectra. The photon emission and absorption rates are proportional to the square of this spectrum and the respective FWHM widths are given by

$$\frac{\Delta\omega_L}{\omega} = \frac{\Delta k'}{k_w} \simeq \frac{1}{N_w}. \tag{2.8}$$

The *classical regime* is defined by the inequality

$$\frac{\omega_a - \omega_e}{\omega} \ll \frac{\Delta\omega_L}{\omega}, \tag{2.9}$$

$$\text{or} \quad \frac{2\hbar\omega}{\gamma mc^2} \ll \frac{1}{N_w}. \tag{2.10}$$

This regime is illustrated in figure 2.2. Laser gain at frequency ω is proportional to the net transition rate for stimulated emission over absorption. Therefore, with the fixed displacement between the respective spectra given by (2.6), we see that gain is proportional to the derivative of the sinc^2 (spontaneous) spectrum. This is the same gain lineshape derived from the fully classical theory. That such a derivation should apply in this case is a consequence of the correspondence principal: while the spectral displacement is proportional to $\hbar$, the transition rate coefficient is inversely proportional to $\hbar$, so laser gain in the classical regime is governed by a fully classical equation (Madey 1971).

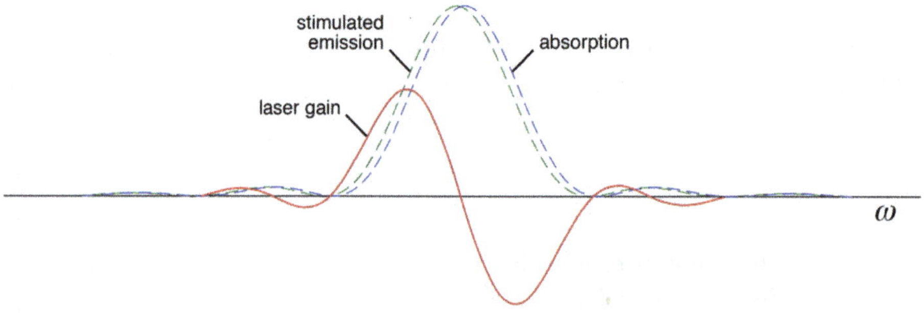

Figure 2.2. The classical FEL regime. Laser gain is proportional to the derivative of the sinc^2 spontaneous spectrum.

2.2 Compton recoil

While (2.9) is an appropriate definition of the classical regime for the reasons discussed above, we need to confirm that the electron–photon interaction can be described using classical electromagnetic fields. This is possible if Compton recoil can be neglected during the interaction; the localization of individual electrons during the bunching process, which leads to classical gain, is then independent of the discrete nature of the quantized electromagnetic field.

To establish this criterion, we require that the net displacement of the electron due to the recoil momentum be much smaller than the characteristic length of the ponderomotive potential—the optical wavelength—during the interaction time. Conservation of longitudinal momentum before and after emission of a photon in the laboratory frame requires

$$\gamma_i m v_i - \gamma_f m v_f = \hbar k \tag{2.11}$$

$$(v_i - v_f)\frac{\mathrm{d}}{\mathrm{d}v}(\gamma m v) = \hbar k \tag{2.12}$$

$$\text{or} \qquad v_i - v_f = \frac{\hbar k}{\gamma^3 m}. \tag{2.13}$$

We can neglect the discrete nature of Compton recoil if $(v_i - v_f)\Delta t_L \ll \lambda$, where $\Delta t_L = \frac{N_W \lambda_W}{c}$ is the interaction time in the undulator:

$$\frac{\hbar k}{\gamma^3 m}\frac{N_w \lambda_w}{c} \ll \lambda \tag{2.14}$$

$$\frac{\hbar\left(2\gamma^2 k_w\right)}{\gamma^3 m}\frac{N_w \lambda_w}{c} \ll \frac{2\pi c}{\omega} \tag{2.15}$$

$$\frac{2\hbar\omega}{\gamma m c^2} \ll \frac{1}{N_w}. \tag{2.16}$$

This inequality corresponds identically to the classical regime defined in equation (2.10).

2.3 Wavepacket spreading

The final justification to employ a fully classical description of the FEL interaction is to neglect the quantum mechanical nature of the electrons, which is possible if wavepacket spreading can be neglected during the interaction time. As with the Compton recoil condition above, this condition preserves the localization of individual electrons during the bunching process leading to classical gain.

Since the standard formula for wavepacket spreading derives from a solution to the non-relativistic Schrödinger equation, we consider wavepacket spreading in the ERF.

For an initially 'unchirped' Gaussian wavepacket of width σ at time $t = 0$, whose probability density is given by $|\psi|^2 \sim \exp[-z^2/2\sigma^2]$, the width σ_t of the wavepacket at a later time t is derived in standard quantum mechanics texts as

$$\sigma_t^2 = \sigma^2 + \frac{\hbar^2 t^2}{4m^2\sigma^2}. \tag{2.17}$$

This expression for σ_t^2 as a function of σ^2 obtains a minimum value (at fixed time t) given by $\sigma_{t;\,\min}^2 = \frac{\hbar}{m}t$. In the ERF, the interaction time is the time for the Lorentz-contracted undulator field to traverse the electron: $\Delta t_L = \frac{N_W \lambda_W}{c\gamma}$. To localize the electron within an optical wavelength, we require that $\sigma_{t;\,\min}$ at time $t = \Delta t_L$ be much smaller than the optical wavelength λ_w/γ in the ERF. Thus, we require

$$\frac{\hbar}{m}\frac{N_w \lambda_w}{c\gamma} \ll \left(\frac{\lambda_w}{\gamma}\right)^2 \tag{2.18}$$

$$\frac{\hbar}{m}\frac{N_w}{c\gamma} \ll \frac{1}{\gamma}\frac{(2\gamma^2\lambda)}{\gamma} \tag{2.19}$$

$$\frac{\hbar\omega}{4\pi\gamma mc^2} \ll \frac{1}{N_w}, \tag{2.20}$$

which, apart from a numerical factor of $\frac{1}{8\pi}$, corresponds identically to the classical regime defined in (2.10). A similar analysis for wavepacket spreading in the transverse direction (Friedman *et al* 1988) leads to the representation of the Compton wavelength $\lambda_C = \frac{h}{mc}$ as the 'normalized emittance' ϵ^n of a single electron. The criterion that the electron trajectories be treated as filamentary during their passage through the undulator is $\epsilon^n/\gamma \ll \lambda$ (see chapter 3).

Contemporary FELs operate well within the classical regime. In the MkIII FEL with $N_w = 47$, $\lambda \sim 3$ μm and $\gamma \sim 90$, the inequality in (2.10) is satisfied by a factor of 10^6, and the criterion $\lambda_C/\gamma \ll \lambda$ is satisfied by a factor of 10^8. We are therefore justified in employing a fully classical analysis, in which the undulator and optical fields are described by the classical Maxwell equations and the electron motion and beam parameters are described by the classical equations of relativistic mechanics.

References

Friedman A, Gover A, Kurizki G, Ruschin S and Yariv A 1988 Spontaneous and stimulated emission from quasifree electrons *Rev. Mod. Phys.* **60** 471–535

Madey J M J 1971 Stimulated emission of bremsstrahlung in a periodic magnetic field *J. Appl. Phys.* **42** 1906–13

Chapter 3

Electron beam dynamics

3.1 Phase space and emittance

The spatial quality of a relativistic electron beam is defined by how well collimated and how tightly focused—simultaneously—the beam can be configured in applications such as linear colliders or FELs. These properties of the electron beam are quantified in the concept of emittance ϵ_x and ϵ_y, defined, respectively, as the coordinate-area occupied by the electrons in $\{x, \theta_x\}$ and $\{y, \theta_y\}$ phase space, with smaller values of the emittance representing a higher quality beam. An example of the $\{x, \theta_x\}$ phase space area occupied by relativistic electrons at the waist of a horizontally focused beam is illustrated in figure 3.1. The $\{y, \theta_y\}$ phase space and emittance ϵ_y are generally independent.

The physical significance of emittance derives from *Liouville's* theorem, which asserts that the volume density of states occupied by the noninteracting, relativistic electrons in six-dimensional phase space $\{\vec{r}, \vec{p}\}$ is a constant of the motion. If the x, y, z motions of the electrons are uncoupled, then the transverse projections of this phase space are independent and the *area* occupied by the electrons in each of the

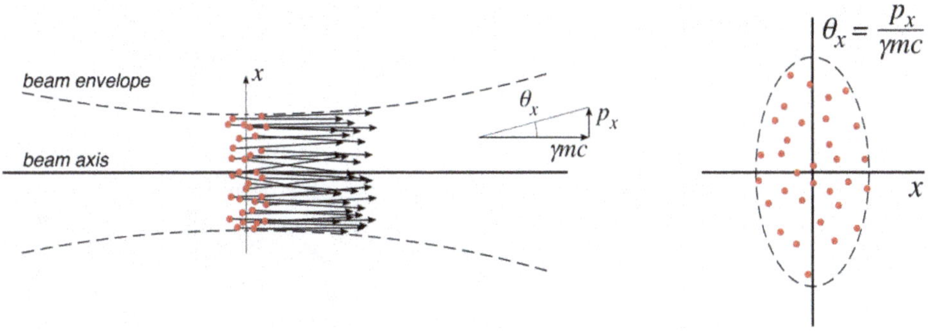

Figure 3.1. Electron positions and velocities in a relativistic beam, and corresponding $\{x, \theta_x\}$ phase space coordinates.

doi:10.1088/978-1-6270-5573-4ch3

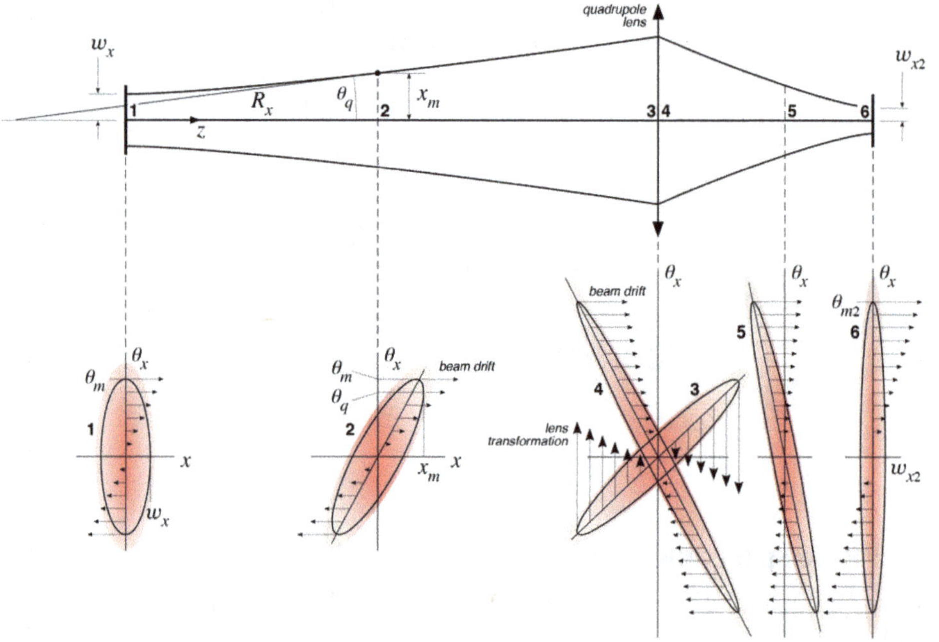

Figure 3.2. Transport of a relativistic electron beam through a lens.

$\left\{x,\, \theta_x\left(=\frac{p_x}{\gamma mc}\right)\right\}$ and $\left\{y,\, \theta_y\left(=\frac{p_y}{\gamma mc}\right)\right\}$ phase spaces is a constant of the motion; the shapes of these phase space regions can change, but their areas and pointwise densities remain invariant. This result is illustrated in figure 3.2, which shows the transport of an electron beam from a waist of radius w_x to a waist of radius $w_{x2} < w_x$ in $\{x, \theta_x\}$ phase space.

To define the emittance, we choose by convention a specified boundary value for the phase space density, such as the $1/e$ boundary indicated by the solid elliptical curves in the phase space distributions in figure 3.2. If the phase space area is represented by an ellipse, as is often the case for well engineered electron beams, then the boundary at the leftmost waist of the beam (where the ellipse is upright) consists of coordinates $\{x, \theta_x\}$ satisfying

$$\frac{x^2}{w_x^2} + \frac{\theta_x^2}{\theta_m^2} = 1. \tag{3.1}$$

The *emittance* ϵ_x is defined as the area enclosed by this boundary and is given by

$$\epsilon_x \equiv \pi w_x \theta_m = \pi w_x \frac{p_m}{\gamma mc} \equiv \frac{\epsilon_x^n}{\gamma}; \qquad \epsilon_x^n \equiv \frac{\pi w_x p_m}{mc}. \tag{3.2}$$

The *normalized emittance* ϵ_x^n is defined as $\epsilon_x^n \equiv \gamma \epsilon_x$. Similar definitions hold for ϵ_y and ϵ_y^n. The significance of normalized emittance is that Liouville's theorem strictly applies to canonical momentum $p_{x,y} = \gamma m v_{x,y}$, not $\theta_{x,y}$. Since longitudinal acceleration in an RF-linac typically conserves the transverse momenta, normalized emittance remains invariant even during acceleration and is independent of the final energy γ.

3.1.1 Beam envelope equation

For the purpose of mode matching the electron beam into the undulator, it is important to know how the radius x_m of the physical beam envelope at the $1/e$ phase space boundary varies as a function of z during transport. The relationship between the beam boundary and the associated phase space distribution is indicated at point 2 in figure 3.2.

To understand our procedure, note that as the beam just starts to propagate away from the waist, the maximum radius x_m will be defined by points of increasing θ_x on the phase space ellipse, just above the x-axis and moving to the right; points closer to the top of the ellipse near θ_m, despite moving to the right with greater velocity, start too far back in x to overtake the points near the x-axis. However, as the beam propagates farther from the waist, the maximum radius x_m will be defined by points on the phase space ellipse closer to the top, since those points move faster and will have had more time to drift ahead of the lower points.

The procedure for deriving the beam envelope equation is to calculate which value of $\theta_x \equiv \theta_q$ on the phase space boundary (see figure 3.2) maximizes the corresponding value of $x \equiv x_m$, for any fixed position z along the beam axis. By reference to (3.1), for a given value of θ_x on the boundary we have

$$x(\theta_x; z) = x(\theta_x; 0) + \theta_x z \tag{3.3}$$

$$= w_x\sqrt{1 - \frac{\theta_x^2}{\theta_m^2}} + \theta_x z, \tag{3.4}$$

where $z = 0$ at the waist. Setting $\frac{\partial x}{\partial \theta_x}\big|_z = 0$ then yields

$$\frac{\theta_q}{\theta_m} = \frac{\theta_m z}{\sqrt{w_x^2 + \theta_m^2 z^2}}; \qquad \sqrt{1 - \frac{\theta_q^2}{\theta_m^2}} = \frac{w_x}{\sqrt{w_x^2 + \theta_m^2 z^2}}, \tag{3.5}$$

for which we find

$$x_m(z) = x(\theta_q; z) = w_x\sqrt{1 + \frac{\theta_m^2 z^2}{w_x^2}}. \tag{3.6}$$

This gives the solution. By further reference to (3.2) this result may be written

$$x_m(z) = w_x\sqrt{1 + \left(\frac{\epsilon_x z}{\pi w_x^2}\right)^2} \equiv w_x\sqrt{1 + \frac{z^2}{\beta_x^2}}, \tag{3.7}$$

where the *beta function* β_x for the horizontal section of the beam is defined by the relations

$$\beta_x \equiv \frac{w_x}{\theta_m} = \frac{\pi w_x^2}{\epsilon_x}, \qquad \text{or} \qquad \pi w_x^2 = \epsilon_x^n \frac{\beta_x}{\gamma}. \tag{3.8}$$

Similar equations hold for the $y_m(z)$ beam envelope and beta function β_y. The beta functions for the electron beam are thus analogous to the Rayleigh range for a Gaussian optical beam (see section 8.1), with $\epsilon \leftrightarrow \lambda$.

The analogous equation for the 'radius of curvature' R_x of the electron beam can also be derived from the above solution. From figure 3.2, the radius R_x at point 2 is given by

$$R_x = \frac{x_m}{\theta_q}. \tag{3.9}$$

Substituting the solution for x_m from (3.6) and the solution for θ_q from the first of (3.5) yields

$$R_x = \frac{w_x^2 + \theta_m^2 z^2}{\theta_m^2 z} = z + \frac{\beta_x^2}{z}, \tag{3.10}$$

where we inserted $\beta_x^2 = w_x^2/\theta_m^2$. An exactly analogous equation holds for R_y.

3.2 Focusing properties of the undulator

Consider a plane-polarized undulator in which the magnetic field on the z-axis ($x = 0$; $y = 0$) is given by

$$\vec{B} = \vec{B}(0, 0, z) = \hat{y}\, B_w \sin k_w z. \tag{3.11}$$

The magnetic field at the off-axis points must satisfy Maxwell's equations. If the poles of the undulator are much wider than the gap between the upper and lower 'jaws' (or magnet arrays), then by symmetry we can impose the functional dependencies $B_x = 0$; $B_y = B_y(y, z)$; $B_z = B_z(y, z)$, and tentatively write

$$B_x = 0 \tag{3.12}$$

$$B_y = B_w f(y) \sin k_w z \tag{3.13}$$

$$B_z = B_w g(y) \cos k_w z. \tag{3.14}$$

Then, from $\nabla \cdot \vec{B} = 0$ we have

$$\frac{\partial B_y}{\partial y} + \frac{\partial B_z}{\partial z} = 0 \tag{3.15}$$

$$B_w f'(y) \sin k_w z - B_w g(y) k_w \sin k_w z = 0 \tag{3.16}$$

$$\Rightarrow \quad f'(y) = k_w\, g(y), \tag{3.17}$$

while from $[\nabla \times \vec{B}]_x = 0$ we have

$$\frac{\partial B_z}{\partial y} - \frac{\partial B_y}{\partial z} = 0 \tag{3.18}$$

$$B_w g'(y) \cos k_w z - B_w f(y) k_w \cos k_w z = 0 \tag{3.19}$$

$$\Rightarrow \quad g'(y) = k_w f(y). \tag{3.20}$$

The solutions to (3.17) and (3.20) consistent with the on-axis field specified in (3.11) are $f(y) = \cosh k_w y$ and $g(y) = \sinh k_w y$, yielding

$$B_x = 0 \tag{3.21}$$

$$B_y = B_w \cosh k_w y \sin k_w z \tag{3.22}$$

$$B_z = B_w \sinh k_w y \cos k_w z. \tag{3.23}$$

To determine the implications of this solution for electron motion, examine the Lorentz force equations which in CGS units read

$$\frac{\mathrm{d}(\gamma m v_x)}{\mathrm{d}t} = -\frac{e}{c}\left(\vec{v} \times \vec{B}\right)_x = -\frac{e}{c}(v_y B_z - v_z B_y) \simeq \frac{e}{c} v_z B_y \tag{3.24}$$

$$\frac{\mathrm{d}(\gamma m v_y)}{\mathrm{d}t} = -\frac{e}{c}\left(\vec{v} \times \vec{B}\right)_y = +\frac{e}{c} v_x B_z \tag{3.25}$$

$$\frac{\mathrm{d}(\gamma m v_z)}{\mathrm{d}t} = -\frac{e}{c}\left(\vec{v} \times \vec{B}\right)_z = -\frac{e}{c} v_x B_y. \tag{3.26}$$

In the $\hat{x}$ component of these equations, the first term in v_y can be neglected relative to the v_z term for most low-emittance beams. The remaining term contains no dependence on v_y and only second order dependence on y through the $\cosh k_w y$ factor in B_y. Therefore, the electron motion in the horizontal plane is substantially independent of both the off-axis dependence of the magnetic field and the motion in the vertical plane. The $\hat{z}$ component of the Lorentz equation is similarly independent of y and v_y.

The $\hat{y}$ component of the Lorentz equation, (3.25), depends on the solution for v_x from (3.24), and also contains a first order dependence on y through the $\sinh k_w y$ factor in B_z. We must solve this equation in conjunction with the solution of (3.24). The solution of (3.24), setting $\cosh k_w y \simeq 1$ in (3.22) together with $v_z \simeq c$ and $z \simeq ct$, is

$$\frac{\mathrm{d}(\gamma m v_x)}{\mathrm{d}t} = \frac{e}{c} v_z B_y = e B_w \sin k_w c t \tag{3.27}$$

$$v_x = -\frac{e B_w}{\gamma m c k_w} \cos k_w c t. \tag{3.28}$$

Equation (3.25) is then written

$$\frac{\mathrm{d}(\gamma m v_y)}{\mathrm{d}t} = \frac{e}{c} v_x B_z = -\frac{e^2 B_w^2}{\gamma m c^2 k_w} \sinh k_w y \cos^2 k_w ct \simeq -\frac{e^2 B_w^2}{\gamma m c^2 k_w} (k_w y) \frac{1}{2}, \quad (3.29)$$

where we approximated $\sinh k_w y \simeq k_w y$ and also took the time-average of the $\cos^2 k_w ct$ factor, appropriate for obtaining the long range behavior of v_y over many periods in the undulator. We thus obtain

$$\frac{\mathrm{d}v_y}{\mathrm{d}t} = -\frac{e^2 B_w^2}{2\gamma^2 m^2 c^2} y, \qquad \text{or} \qquad \frac{\mathrm{d}^2 y}{\mathrm{d}t^2} = -\kappa^2 y, \quad (3.30)$$

where the constant κ^2 is

$$\kappa^2 = \frac{e^2 B_w^2}{2\gamma^2 m^2 c^2} \equiv \frac{\hat{K}^2 k_w^2 c^2}{\gamma^2}, \quad (3.31)$$

$$\text{and we define} \quad \hat{K} \equiv \frac{e\hat{B}_w}{mc^2 k_w} = \frac{e\hat{B}_w \lambda_w}{2\pi mc^2} \quad (3.32)$$

as the rms *undulator parameter* proportional to the rms undulator magnetic field $\hat{B}_w = B_w/\sqrt{2}$. Upon dividing both sides of (3.30) by c^2, we obtain the desired result

$$\frac{\mathrm{d}^2 y}{\mathrm{d}z^2} = -\alpha^2 y, \qquad \text{where} \qquad \alpha = \frac{\hat{K} k_w}{\gamma}. \quad (3.33)$$

The electrons exhibit stable oscillatory orbits in the vertical plane. These so-called *betatron oscillations* are a consequence of the focusing effect due to the off-axis dependence of the undulator magnetic field.

3.3 Matching into the FEL

To further quantify the vertical focusing properties of the undulator, consider the general solution of (3.33),

$$y(z) = y(0) \cos \alpha z + \frac{y'(0)}{\alpha} \sin \alpha z \quad (3.34)$$

$$y'(z) = -\alpha y(0) \sin \alpha z + y'(0) \cos \alpha z. \quad (3.35)$$

First, we note that the spatial period of the betatron oscillations is

$$\lambda_\beta = \frac{2\pi}{\alpha} = \frac{2\pi\gamma}{k_w \hat{K}} = \frac{\lambda_w \gamma}{\hat{K}}, \quad (3.36)$$

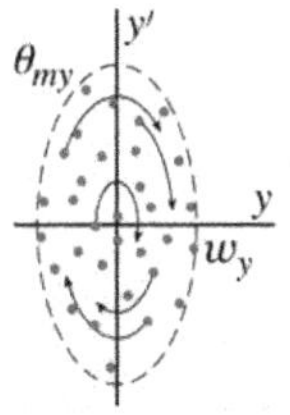

Figure 3.3. Phase space motion of a matched electron beam in the B-field plane of a plane-polarized undulator magnet.

and the number of betatron oscillations in a single pass through the undulator is

$$n_{\text{osc}} = \frac{L_w}{\lambda_\beta} = \frac{N_w \hat{K}}{\gamma}. \tag{3.37}$$

For $\hat{K}^2 = 1.2$ and $\gamma = 90$ this yields about 0.6 oscillations in the Mark III FEL ($N_w = 47$). Second, we see that the electrons execute elliptical orbits in $\{y, \theta_y\}$ phase space, with an aspect ratio (choosing $y'(0) = 0$) of

$$\frac{y_{\max}}{y'_{\max}} = \frac{y(0)}{\alpha\, y(0)} = \frac{1}{\alpha}. \tag{3.38}$$

If we want the vertical section of the electron beam to propagate as a collimated beam of fixed radius, then this aspect ratio must match the aspect ratio of the injected phase space distribution, as shown in figure 3.3 ($y' \equiv \theta_y$). The electron beam in that case must enter the undulator at a waist, with β_y satisfying

$$\beta_y \equiv \frac{w_y}{\theta_{my}} = \frac{1}{\alpha} \equiv \frac{\gamma}{\hat{K} k_w}. \tag{3.39}$$

This result can also be obtained another way. Equations (3.34, 3.35) can be written in differential matrix form to order dz as

$$\begin{bmatrix} y(z + \mathrm{d}z) \\ y'(z + \mathrm{d}z) \end{bmatrix} = \begin{bmatrix} 1 & \mathrm{d}z \\ -\alpha^2 \mathrm{d}z & 1 \end{bmatrix} \begin{bmatrix} y(z) \\ y'(z) \end{bmatrix}, \tag{3.40}$$

which describes propagation of the beam over an increment dz. This is the matrix equation for propagation through a lens of thickness dz and focal length

$$f = \frac{1}{\alpha^2 \mathrm{d}z}, \tag{3.41}$$

analogous to the result from the matrix theory of ray propagation in geometric optics. If we want the vertical section of the electron beam to propagate as a collimated beam of fixed radius, then the incident electron beam must be configured so that the vertical focusing effect of the undulator balances the natural property of the beam to 'diffract'. This in turn requires that the 'radius of curvature' R_y developed by the beam over distance $\mathrm{d}z$ from the waist be equal to the focal length of the undulator over the corresponding length $\mathrm{d}z$. By reference to (3.10), we require

$$R_y\left(= \mathrm{d}z + \frac{\beta_y^2}{\mathrm{d}z} = \frac{\beta_y^2}{\mathrm{d}z}\right) = f\left(= \frac{1}{\alpha^2 \mathrm{d}z}\right), \qquad \text{or} \qquad \beta_y = \frac{1}{\alpha}, \tag{3.42}$$

as stated in (3.39). The corresponding radius w_y of the electron beam is obtained from

$$\pi w_y^2 = \epsilon_y \beta_y = \frac{\epsilon_y^n}{\hat{K} k_w}. \tag{3.43}$$

For Mark III FEL parameters of $\epsilon_y^n = 4\pi$ mm·mrad, $\hat{K}^2 = 1.2$, and $\lambda_w = \frac{2\pi}{k_w} = 2.3$ cm, we find $w_y = 116$ µm. This is much smaller than typical optical beam radii ($w_0 \sim 700$ µm or so), and induces only a slight reduction in the small-signal gain of the laser (section 10.1).

The horizontal section of the electron beam is not affected by the undulator magnetic field and propagates as a free-space beam. The procedure to maximize the small-signal gain is to ensure that the horizontal radius of the electron beam remains much smaller than the optical radius over the length of the undulator. Since the spatial envelopes of both beams are governed by formulas of precisely the same form, this condition is best achieved by matching the corresponding 'Rayleigh ranges', so that

$$\beta_x = z_R, \qquad \text{or} \qquad \frac{\pi w_x^2}{\epsilon_x} = \frac{\pi w_0^2}{\lambda}. \tag{3.44}$$

The condition that $w_x^2 \ll w_0^2$ is therefore equivalent, in this mode-matched configuration, to

$$\epsilon_x \ll \lambda, \qquad \text{or} \qquad \frac{\epsilon_x^n}{\gamma} \ll \lambda. \tag{3.45}$$

The realization of an FEL thus imposes stringent physical limitations on the emittance of the electron beam. For typical Mark III FEL parameters of $\epsilon_x^n = 8\pi$ mm·mrad, $\gamma = 90$, and $\lambda = 3$ µm, we have

$$\frac{\epsilon_x^n}{\gamma} = 0.093\,\lambda, \tag{3.46}$$

so this condition is well satisfied.

Brau (1990) obtains the same results derived in this chapter starting from a Hamiltonian analysis of the electron motion and includes a thorough discussion of Liouville's theorem and its relevance to emittance. The subsequent development of the FEL coupled equations of motion in this text will assume an ideal, filamentary electron beam constrained to travel along the z-axis of the undulator. We will include the transverse structure of the electron and optical beams in chapter 8 and quantify the effects of emittance in chapter 10.

Reference

Brau C A 1990 *Free-Electron Lasers* (Boston MA: Academic)

Chapter 4

Undulator trajectories

4.1 Transverse motion

Electron motion in the undulator is governed by the momentum (spatial) components of the Lorentz equation,

$$\frac{\mathrm{d}(\gamma\vec{\beta})}{\mathrm{d}t} = -\frac{e}{mc}\left[\vec{E} + \vec{\beta} \times \vec{B}\right]. \tag{4.1}$$

With the exception of chapter 13, we consider only plane-polarized undulators in this text, for which the on-axis undulator field $\vec{B}_w$ and radiation fields $\vec{E}_r$, $\vec{B}_r$ defining the coordinate system are given by

$$\vec{B}_w = \hat{y}\, B_w \sin k_w z \tag{4.2}$$

$$\vec{E}_r = \hat{x}\ |E|\cos(fkz - f\omega t + \phi) \tag{4.3}$$

$$\vec{B}_r = \hat{y}\ |B|\cos(fkz - f\omega t + \phi). \tag{4.4}$$

We explicitly omit the interparticle Coulomb field derived from the scalar potential Φ, as space charge forces fall off as $1/\gamma^2$ and can thus be neglected in relativistic beams for sufficiently small current densities. Here, $|E| = |B|$ are in CGS units and we allow for the possibility of driving the laser on some harmonic f of the fundamental radiation frequency ω (chapter 12). Typically, FEL oscillation occurs on the fundamental $f = 1$. With ψ denoting the full phase of the radiation fields, the x-component of (4.1) is

$$\frac{\mathrm{d}(\gamma\beta_x)}{\mathrm{d}t} = -\frac{e}{mc}\left[|E|\cos\psi + (-\beta_z\,|B|\cos\psi - \beta_z B_w \sin k_w z)\right]. \tag{4.5}$$

(Of course, $\beta_x = v_x/c$, *not* the 'Rayleigh parameter' of the beam.) We see that the $\vec{E}_r$ and $\vec{B}_r$ radiation forces cancel to order $1 - \beta_z \sim 1/2\gamma^2$, so we neglect them in this

doi:10.1088/978-1-6270-5573-4ch4

analysis, therefore effectively calculating the electron trajectories in the absence of radiation. (The effects of radiation will be developed through the temporal component of the Lorentz equation in chapter 6; the use in that analysis of the trajectories derived in this section will be sufficient to calculate the effects of radiation to order $1/\gamma^2$.) In this approximation (4.5) becomes

$$\frac{d(\gamma\beta_x)}{dt} = +\frac{e}{mc}\beta_z B_w \sin k_w z \tag{4.6}$$

$$v_z\frac{d(\gamma\beta_x)}{dz} = \frac{eB_w}{mc^2} v_z \sin k_w z \tag{4.7}$$

$$\beta_x = -\frac{K}{\gamma}\cos k_w z, \qquad \text{where} \qquad K \equiv \frac{eB_w}{mc^2 k_w} = \frac{eB_w\lambda_w}{2\pi mc^2}. \tag{4.8}$$

The parameter K is known as the *undulator parameter* and is identical to the normalized vector potential of the undulator field. Its rms value $\hat{K} = K/\sqrt{2}$ is also denoted a_w. It is one of the most important parameters in FEL physics because it provides a fundamental measure of the strength of the FEL interaction.

Since β_x equals the angle of propagation of the electrons relative to the undulator axis, the maximum angle is $\theta_{max} = K/\gamma$. The transverse oscillations in the electron trajectory are obtained by integrating the solution for β_x in (4.8),

$$x(t) = \int^t v_x \, dt' = -\int^t \frac{cK}{\gamma}\cos(k_w v_z t')dt' = -\frac{K}{\gamma k_w}\sin(k_w ct), \tag{4.9}$$

where we employed the approximation $v_z \simeq c$.

4.2 Longitudinal motion

Instead of extracting the z-component from the Lorentz equation explicitly, we can use the relation

$$\frac{1}{\gamma^2} = 1 - \left(\beta_x^2 + \beta_y^2 + \beta_z^2\right). \tag{4.10}$$

Inserting $\beta_y = 0$ and the solution for β_x calculated above, we obtain directly

$$\beta_z^2 = 1 - \frac{1}{\gamma^2} - \frac{K^2}{\gamma^2}\cos^2 k_w z \tag{4.11}$$

$$= 1 - \frac{1}{\gamma^2}\left[1 + \hat{K}^2 + \hat{K}^2\cos 2k_w z\right], \tag{4.12}$$

where we substituted $\hat{K} = K/\sqrt{2}$. Expanding the square root to extract β_z, we obtain

$$\beta_z = 1 - \frac{1 + \hat{K}^2}{2\gamma^2} - \frac{\hat{K}^2}{2\gamma^2}\cos 2k_w z + O\left(\frac{1}{\gamma^4}\right) \tag{4.13}$$

or

$$\beta_z = \bar{\beta}_z - \frac{\hat{K}^2}{2\gamma^2}\cos 2k_w z, \tag{4.14}$$

where

$$\bar{\beta}_z \equiv 1 - \frac{1 + \hat{K}^2}{2\gamma^2} \tag{4.15}$$

is the mean velocity of the electrons along the undulator. We thus identify this velocity with the velocity of the ERF. Substitution of this expression for $\bar{\beta}_z$ into (1.5) yields the FEL *resonance condition*, giving the wavelength of FEL spontaneous emission as a function of the electron energy and undulator magnetic field,

$$\lambda = \frac{\lambda_w}{2\gamma^2}\left(1 + \hat{K}^2\right). \tag{4.16}$$

Practically, FELs constructed with NdFe permanent magnets can achieve values of $\hat{K}^2$ approaching 1.5 and usable laser gains can be obtained with values of $\hat{K}^2$ as low as 0.3 or so. Therefore, such FELs can be tuned continuously in laser wavelength by almost a factor of two simply by varying the strength of the magnetic field (usually by varying the physical separation between the upper and lower jaws of the undulator magnet).

To obtain the solution for $z(t)$ as a function of time, which we will need in our derivation of the FEL equations of motion in the plane-polarized undulator, we use the solution from (4.14) to develop a perturbation expansion to order $1/\gamma^2$ as follows:

$$z(t) = \bar{\beta}_z ct - \frac{\hat{K}^2 c}{2\gamma^2}\int^t \cos\left(2k_w z(t')\right)\mathrm{d}t' \tag{4.17}$$

$$= \bar{\beta}_z ct - \frac{\hat{K}^2 c}{2\gamma^2}\int^t \cos\left[2k_w\left(\bar{\beta}_z ct' - \frac{\hat{K}^2 c}{2\gamma^2}\int^{t'}(\ldots)\mathrm{d}t''\right)\right]\mathrm{d}t' \tag{4.18}$$

$$= \bar{\beta}_z ct - \frac{\hat{K}^2 c}{2\gamma^2}\int^t \cos(2k_w\bar{\beta}_z ct')\mathrm{d}t' + O\left(\frac{1}{\gamma^4}\right) \tag{4.19}$$

$$= \bar{\beta}_z ct - \frac{\hat{K}^2}{4k_w\gamma^2}\sin(2k_w ct), \tag{4.20}$$

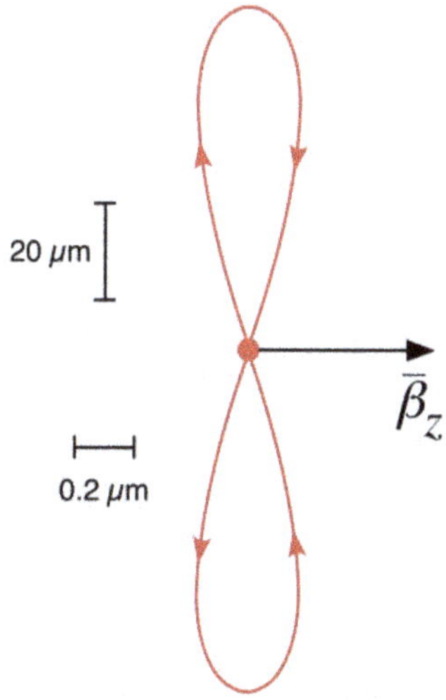

Figure 4.1. Electron motion in the ERF in a plane-polarized undulator for $\hat{K}^2 = 1.2$; $\gamma = 80$; $k_w = 2.73\ \mathrm{cm}^{-1}$. The maximum transverse and longitudinal displacements are ±71 μm and ±0.17 μm respectively.

where we employed $\bar{\beta}_z \simeq 1$ in the second term. We now use the expression $1 - \bar{\beta}_z = \frac{k_w}{k} = (1 + \hat{K}^2)/2\gamma^2$ to substitute for $k_w\gamma^2$ in the denominator to obtain

$$z(t) = \bar{\beta}_z ct - \frac{\eta}{k}\sin(2k_w ct); \qquad \eta \equiv \frac{\hat{K}^2}{2\left(1 + \hat{K}^2\right)}. \tag{4.21}$$

The parameter η quantifies the emission of radiation into higher harmonics. Superposition of the solutions for $x(t)$ and $z(t)$ yields the characteristic 'figure-8' motion of the electron in the ERF (traveling at speed $\bar{\beta}_z c$), as illustrated in figure 4.1.

Chapter 5

Spontaneous emission

5.1 Spectral lineshape

Consider a single electron traveling though a uniform undulator magnet containing N_w periods of wavelength λ_w (figure 5.1). From the discussion of Thomson scattering in section (1.2), it is clear that the fundamental optical radiation ($f = 1$) will appear in a pulse that has the same rectangular profile as the undulator magnetic field and contain N_w optical cycles of wavelength λ. The complex analytic waveform of this radiation is

$$E(t) = \begin{cases} E_0 e^{-i\omega_r t} & \text{for } |t| \leqslant \dfrac{\tau_p}{2}, \\ 0 & \text{otherwise} \end{cases} \tag{5.1}$$

where $\tau_p = N_w\lambda/c$ is the optical pulse duration and ω_r is the *resonant frequency*

$$\omega_r = \frac{2\pi c}{\lambda} = 2\pi c\,\frac{2\gamma^2}{\lambda_w\left(1 + \hat{K}^2\right)}. \tag{5.2}$$

The optical spontaneous spectrum is the Fourier transform of $E(t)$,

$$\tilde{E}(\omega) = \int_{-\frac{\tau_p}{2}}^{+\frac{\tau_p}{2}} E_0 e^{-i\omega_r t} e^{i\omega t} dt = \frac{2E_0}{\omega - \omega_r} \sin\left[(\omega - \omega_r)\frac{N_w\lambda}{2c}\right]$$

$$= \frac{2E_0}{\omega - \omega_r} \sin\left[\frac{\pi N_w}{\omega_r}(\omega - \omega_r)\right], \tag{5.3}$$

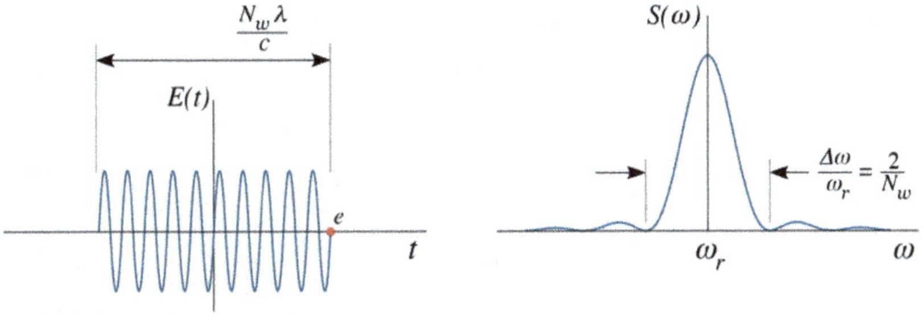

Figure 5.1. Optical field and power spectrum of spontaneous radiation in a uniform undulator with N_w periods ($f = 1$).

with power spectral density $$S(\omega) = |\tilde{E}(\omega)|^2 = S_0 \,\mathrm{sinc}^2\left[\frac{N_w}{\omega_r}(\omega - \omega_r)\right], \tag{5.4}$$

where $\mathrm{sinc}(x) \equiv [\sin \pi x]/[\pi x]$ (figure 5.1). The spontaneous spectrum from an incoherent beam of monoenergetic electrons has the same lineshape. The fractional width of the spontaneous spectrum for harmonic f is narrower by a factor of f, because the pulse length τ_p is the same while the frequency is increased by the factor f.

5.2 Spontaneous power (weak undulator fields)

To calculate the energy emitted by a single electron in its passage through the undulator, we employ Jackson (1975; equation (14.36)), which integrates the radial component of the Poynting vector during the interaction,

$$\Delta U = \int_{t_1=t_1'+R(t_1')/c}^{t_2=t_2'+R(t_2')/c} [\vec{S}\cdot\hat{n}]_{\mathrm{ret}} \Delta A \,\mathrm{d}t = \int_{t_1'}^{t_2'} [\vec{S}\cdot\hat{n}]\Delta A \frac{\mathrm{d}t}{\mathrm{d}t'}\,\mathrm{d}t'. \tag{5.5}$$

We first consider the case of weak undulator fields, for which the particle velocity is $\vec{\beta} \simeq \beta\hat{z}$. There are two methods by which we can calculate the energy: the first '=' sign in the above equation refers to integration of the *optical* power in the far field, which spans a time interval $t_2 - t_1 = N_w\lambda/c$ equal to the optical pulse duration at the receiver. The second '=' sign in the above equation refers to integration of the power *emitted by the electron*, spanning a time interval $t_2' - t_1' = N_w\lambda_w/v_z$ equal to the transit time through the undulator. The element $\mathrm{d}t$ is the time interval measured by a laboratory clock located at the receiver and the element $\mathrm{d}t'$ is the retarded time interval measured by a series of synchronized laboratory clocks located along the trajectory of the particle. Since both methods must yield the same energy, the respective integrands differ by a factor (taking $\hat{n} = \hat{z}$) of

$$\frac{\mathrm{d}t}{\mathrm{d}t'} = 1 - \vec{\beta}\cdot\hat{n} = 1 - \beta = \frac{\lambda}{\lambda_w}. \tag{5.6}$$

The same relation between the integration intervals, $\Delta t = \Delta t'(1 - \beta)$, can be obtained directly from figure 5.2.

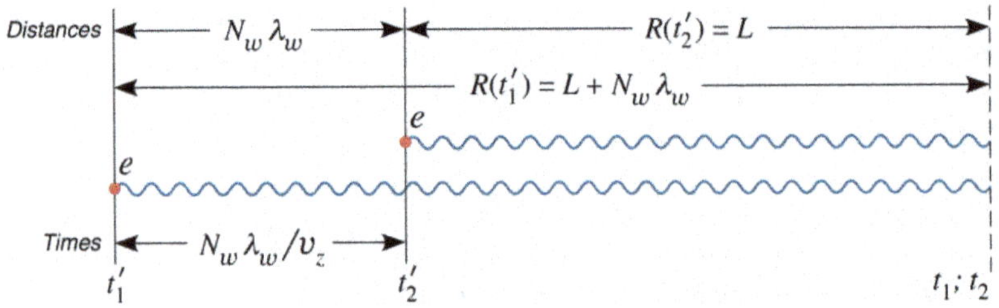

Figure 5.2. Calculation of retarded times for undulator radiation: $(t_2 - t_1) = (t_2' - t_1')(1 - \beta)$.

The power radiated per unit solid angle by the electron during its transit through the undulator is given by Jackson (1975; equation (14.38)),

$$\frac{\mathrm{d}P_e(t')}{\mathrm{d}\Omega} = R^2[\vec{S}\cdot\hat{n}]\frac{\mathrm{d}t}{\mathrm{d}t'} = \frac{e^2}{4\pi c}\frac{\left|\hat{n}\times\left\{(\hat{n}-\vec{\beta})\times\dot{\vec{\beta}}\right\}\right|^2}{\left(1-\vec{\beta}\cdot\hat{n}\right)^5}, \tag{5.7}$$

where the subscript e on the lhs here indicates 'emitted' power, not optical power.

To obtain expressions for $\vec{\beta}$ and $\dot{\vec{\beta}}$ to lowest order in $1/\gamma$ in weak undulator fields $K^2 \ll 1$, we take $\vec{\beta} = \beta_z\hat{z} = \beta\hat{z}$, and

$$\dot{\vec{\beta}} = \dot{\beta}_x\hat{x} = \frac{eB_w}{\gamma mc}\sin(k_w z)\hat{x} = \frac{k_w Kc}{\gamma}\sin(k_w ct')\hat{x}, \tag{5.8}$$

where we used (4.6) with $\beta_z \simeq 1$ and $z = ct'$. Taking $\hat{n} = \hat{z}$, we then calculate for on-axis emission

$$\left\langle\frac{\mathrm{d}P_e}{\mathrm{d}\Omega}\right\rangle = \frac{e^2}{4\pi c}\left\langle\frac{\left|\hat{z}\times\left\{(\hat{z}-\beta\hat{z})\times\dot{\beta}_x\hat{x}\right\}\right|^2}{(1-\beta\hat{z}\cdot\hat{z})^5}\right\rangle_{t'} \tag{5.9}$$

$$= \frac{e^2}{4\pi c}\frac{(1-\beta)^2}{(1-\beta)^5}\langle\dot{\beta}_x^2\rangle_{t'} \tag{5.10}$$

$$= \frac{e^2}{4\pi c}\frac{1}{(1/2\gamma^2)^3}\frac{k_w^2 K^2c^2}{\gamma^2}\langle\sin^2 k_w ct'\rangle_{t'} \tag{5.11}$$

$$= \left(\frac{e^2}{mc^2}\right)\frac{mc^2}{\pi c}K^2k_w^2\gamma^4c^2 \tag{5.12}$$

$$= \frac{r_0}{\pi}mc^3K^2k_w^2\gamma^4, \tag{5.13}$$

where r_0 is the classical radius of the electron and we used $\langle \sin^2 k_w ct' \rangle_{t'} = 1/2$ in the fourth line. This result agrees with the on-axis value of the angular distribution obtained by Hofmann (1986) for small K, who calculates the Thomson-scattered electric and magnetic fields in the ERF, transforms the fields into the lab frame and scales the resulting optical power by λ/λ_w to convert from optical power to 'emitted' power, thus obtaining

$$\left\langle \frac{\mathrm{d}P_e}{\mathrm{d}\Omega} \right\rangle = \frac{r_0 mc^3 K^2 k_w^2 \gamma^4}{\pi(1+\gamma^2\theta^2)^5} \left[\left(1 - \gamma^2\theta^2 \cos 2\phi\right)^2 + \left(\gamma^2\theta^2 \sin 2\phi\right)^2 \right]. \tag{5.14}$$

The same angular distribution can be obtained from (5.7) by substituting $\hat{n} = (\sin\theta \cos\phi, \sin\theta \sin\phi, \cos\theta)$. The total energy radiated per unit solid angle by the electron during its transit through the undulator is then

$$\frac{\mathrm{d}U}{\mathrm{d}\Omega} = \frac{N_w \lambda_w}{c} \cdot \left\langle \frac{\mathrm{d}P_e}{\mathrm{d}\Omega} \right\rangle, \tag{5.15}$$

where $N_w \lambda_w / c$ is the interaction time for propagation through the undulator at speed $v_z \simeq c$.

If the electron radiates into free space, as in synchrotron light sources, the total emitted energy is calculated by integrating (5.15) over elements of solid angle $\mathrm{d}\Omega = \sin\theta \mathrm{d}\theta \mathrm{d}\phi$. Since the angular distribution is sharply peaked in the forward direction within a cone of angular width $\sim 1/\gamma$, the integration is made tractable by substituting $\mathrm{d}\Omega = \theta \mathrm{d}\theta \mathrm{d}\phi$. Integration over the angle-dependent factors in (5.14) then yields

$$\begin{aligned} &\int_0^{2\pi} \mathrm{d}\phi \int_0^{\pi} \mathrm{d}\theta\, \theta \frac{\left(1 - \gamma^2\theta^2 \cos 2\phi\right)^2 + \left(\gamma^2\theta^2 \sin 2\phi\right)^2}{(1+\gamma^2\theta^2)^5} \\ &= 2\pi \int_0^{\pi} \mathrm{d}\theta\, \theta \frac{1 + \gamma^4\theta^4}{(1+\gamma^2\theta^2)^5} \\ &= \frac{\pi}{\gamma^2} \int_0^{\infty} \mathrm{d}x \frac{1 + x^2}{(1+x)^5} = \frac{\pi}{3\gamma^2}. \end{aligned} \tag{5.16}$$

The total energy U radiated by a single particle into free space is then

$$U_{\text{free space}} = \frac{N_w \lambda_w}{c} \cdot P_e, \qquad \text{where} \qquad P_e = \frac{r_0}{3} mc^3 K^2 k_w^2 \gamma^2. \tag{5.17}$$

If the electron emits into an optical resonator, as in an FEL, then the quantity of interest is the optical power that couples directly to the spatial modes defined by the boundary conditions imposed by the cavity mirrors. This power is relevant for the buildup of laser oscillation in each of the modes. Typical FELs employ long, narrow undulator vacuum bores that impose severe diffractive losses on all but the lowest order transverse mode and thus in many systems only the lowest order mode is supported.

Consider emission into the fundamental Gaussian mode of an optical resonator. For typical FEL resonator geometries with mode radius w_0 at the waist, the divergence half-angle $\theta_{1/e} = \lambda/\pi w_0$ in the far field satisfies $\theta_{1/e} \lesssim 1$ mrad. Thus, for typical electron energies of $\gamma \lesssim 100$ or so we have $\gamma^2\theta^2 \lesssim 0.01$, and we may take the paraxial angular distribution in (5.14) to be uniform over the cross section of the mode. For a spherical wave of uniform power density, the solid angle subtended by the lowest order Gaussian mode (from a standard calculation in Gaussian mode analysis) is

$$\Delta\Omega = 2\frac{\lambda^2}{\pi w_0^2} \tag{5.18}$$

which, for a paraxial beam, is insensitive to small displacements of the point source within the confocal region. The total energy emitted by a single particle into the fundamental mode of the resonator for $K^2 \ll 1$ is then

$$U_{\mathrm{mode}} = \frac{r_0}{\pi} mc^3 K^2 k_w^2 \gamma^4 \frac{N_w \lambda_w}{c} \cdot \frac{2\lambda^2}{\pi w_0^2}; \qquad K \ll 1. \tag{5.19}$$

The corresponding optical power is obtained by dividing this result by the optical pulse duration $N_w\lambda/c$. If we want the average instantaneous optical power emitted into the optical mode by a long electron bunch of duration $\tau_b \gg N_w\lambda/c$, then for incoherent emission we have

$$\langle P_{\mathrm{opt}} \rangle = \frac{\text{Total radiated energy}}{\text{Bunch duration}} = \frac{N_e U_{\mathrm{mode}}}{\tau_b}, \tag{5.20}$$

where N_e is the total number of electrons in the bunch.

5.3 Spontaneous power (strong undulator fields)

The radiation emitted by a single electron in a large magnetic field ($K \gtrsim 1$) is accompanied by substantial emission of higher harmonics of the fundamental frequency. The appearance of harmonics can be understood by recalling that the maximum propagation angle of the electron is $\theta_{\mathrm{max}} = K/\gamma$, while the radiation cone subtends an angle $\theta_{\mathrm{rad}} \sim 1/\gamma$. Consequently, when $K \gtrsim 1$ we have $\theta_{\mathrm{max}} \gtrsim \theta_{\mathrm{rad}}$, and the radiation in the far field at all angles exhibits a manifest deviation from pure sinusoidal oscillation (the so-called 'searchlight effect').

The even harmonics of the fundamental frequency exhibit zero on-axis emission. Hofmann (1986) calculates the following result for on-axis emission by a single particle into the fth odd harmonic:

$$\left\langle \frac{dP_e}{d\Omega} \right\rangle_f = \frac{3f^2\gamma^2 P_e}{\pi\left(1+\hat{K}^2\right)^3} \left[J_{\frac{f-1}{2}}(f\eta) - J_{\frac{f+1}{2}}(f\eta) \right]^2; \qquad f = 1, 3, 5, \ldots \tag{5.21}$$

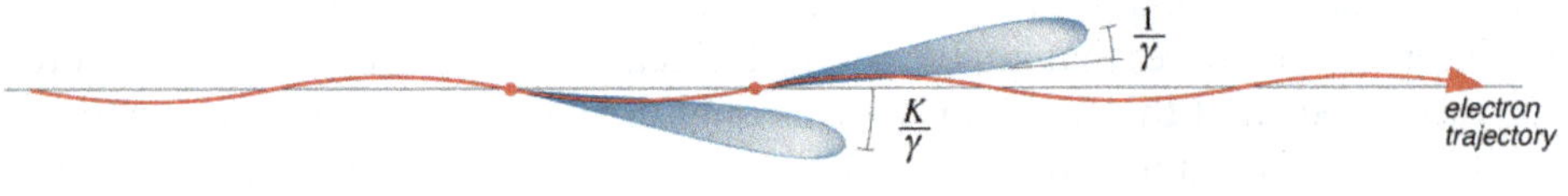

Figure 5.3. The origin of harmonic emission for $K \gtrsim 1$.

where $J_n(x)$ is the nth-order Bessel function of the zeroth kind, η is defined in (4.21) and P_e is the total power emitted into free space by a single particle in the weak field limit, (5.17). To calculate the harmonic power radiated into the fundamental resonator mode, the above result for $\langle dP_e/d\Omega\rangle_f$ should be multiplied by the expression for $\Delta\Omega$ from (5.18), where λ and w_0 are those of the harmonic mode. The corresponding energy is then obtained by multiplying the power by the interaction time $N_w\lambda_w/c$.

References

Hofmann A 1986 Theory of synchrotron radiation *Stanford Synchrotron Radiation Laboratory, (SSRL)* ACD-NOTE 38

Jackson J D 1975 *Classical Electrodynamics* 2nd edn (New York: Wiley)

Chapter 6

Effect of the optical field on electron motion

6.1 The Lorentz equation

The derivation of the FEL pendulum equation in this chapter is based on the analysis of Colson (1981). To begin, we note that the time rate of change of electron energy is governed by the temporal component of the Lorentz equation, which in CGS units reads

$$\frac{\mathrm{d}\gamma}{\mathrm{d}t} = -\frac{e}{mc}\vec{\beta} \cdot \vec{E}_r = -\frac{e}{mc}\beta_x E_{r,x}, \tag{6.1}$$

where

$$\vec{E}_r = \hat{x}\,|E|\cos(fkz - f\omega t + \phi) \tag{6.2}$$

is the co-propagating optical field in the lab frame, $E \equiv |E|\mathrm{e}^{\mathrm{i}\phi}$ is the complex envelope and β_x is given by (4.8). The interparticle Coulomb field is again omitted for the reasons discussed in section 4.1. (Note that our convention $\mathcal{Re}\{\mathrm{e}^{\mathrm{i}(fkz-f\omega t+\phi)}\}$ in (6.2) defines our sign of the optical phase ϕ. All complex quantities appearing in this text would be conjugated if we had chosen $\mathcal{Re}\{\mathrm{e}^{\mathrm{i}(f\omega t-fkz+\phi)}\}$.) The Lorentz equation is thus written

$$\frac{\mathrm{d}\gamma}{\mathrm{d}t} = +\frac{eK|E|}{\gamma mc}\cos(k_w z)\cos(fkz - f\omega t + \phi) \tag{6.3}$$

$$= +\frac{eK|E|}{2\gamma mc}[\cos(k_w z + fkz - f\omega t + \phi) + \cos(-k_w z + fkz - f\omega t + \phi)]. \tag{6.4}$$

doi:10.1088/978-1-6270-5573-4ch6

We now insert the time-dependent solution for $z(t)$ from (4.21) and add and subtract $fk_w\bar{z}$ in each argument (where $\bar{z} = \bar{\beta}_z ct$) to obtain

$$\frac{d\gamma}{dt} = \frac{eK|E|}{2\gamma mc}\Bigg[\cos\Bigg(+k_w\bar{z} - \frac{k_w\eta}{k}\sin(2k_wct) + fk\bar{z}$$

$$- f\eta\sin(2k_wct) - f\omega t + \phi + fk_w\bar{z} - fk_w\bar{z}\Bigg)$$

$$+\cos\Bigg(-k_w\bar{z} + \frac{k_w\eta}{k}\sin(2k_wct) + fk\bar{z}$$

$$- f\eta\sin(2k_wct) - f\omega t + \phi + fk_w\bar{z} - fk_w\bar{z}\Bigg)\Bigg]. \tag{6.5}$$

Dropping terms in $k_w/k \sim 1/\gamma^2$ in each argument and regrouping the remaining terms yields

$$\frac{d\gamma}{dt} = \frac{eK|E|}{2\gamma mc}\big[\cos(f[(k + k_w)\bar{z} - \omega t] + \phi - (f - 1)k_wct - f\eta\sin(2k_wct))$$

$$+ \cos(f[(k + k_w)\bar{z} - \omega t] + \phi - (f + 1)k_wct - f\eta\sin(2k_wct))\big]. \tag{6.6}$$

Following section (1.2) of the introduction, we define the *electron phase* ξ in the ponderomotive potential and its time derivative, the *phase velocity* $\nu = d\xi/d\tau$,

$$\xi = (k + k_w)\bar{z} - \omega t \tag{6.7}$$

$$\nu = L_w\big[(k + k_w)\bar{\beta}_z - k\big], \tag{6.8}$$

where the dimensionless time $\tau = ct/L_w$ varies from 0 to 1 along the undulator of length $L_w = N_w\lambda_w$. The parameters (ξ, ν) are independent phase space coordinates that track the evolution of the electrons as they drift in the ponderomotive potential in the presence of a co-propagating optical field. Equation (6.6) is then

$$\frac{d\gamma}{dt} = \frac{eK|E|}{2\gamma mc}[\cos(f\xi + \phi - (f - 1)k_wct - f\eta\sin(2k_wct))$$

$$+ \cos(f\xi + \phi - (f + 1)k_wct - f\eta\sin(2k_wct))]$$

$$= \frac{e\hat{K}|\hat{E}|}{\gamma mc}\mathcal{Re}\left\{e^{i(f\xi+\phi)}e^{-i(f-1)k_wct}e^{-if\eta\sin(2k_wct)} + e^{i(f\xi+\phi)}e^{-i(f+1)k_wct}e^{-if\eta\sin(2k_wct)}\right\}, \tag{6.9}$$

where we introduced the rms parameters $\hat{K}$ and $|\hat{E}|$. We now insert the Bessel function identity

$$e^{ix\sin y} = \sum_{n=-\infty}^{\infty} J_n(x)e^{iny}, \tag{6.10}$$

yielding

$$\frac{d\gamma}{dt} = \frac{e\hat{K}|\hat{E}|}{\gamma mc}\mathcal{R}e\left\{e^{i(f\xi+\phi)}\left[e^{-i(f-1)k_w ct}\sum_n J_n(f\eta)e^{-in2k_w ct} + e^{-i(f+1)k_w ct}\sum_n J_n(f\eta)e^{-in2k_w ct}\right]\right\}. \tag{6.11}$$

Since we are interested in the slow evolution of the electron energy γ over many magnet periods, we *average* the above equation over the magnet period $c\Delta t = \lambda_w$. This procedure yields only one surviving dc term from each sum, $2n = -(f-1)$ and $2n = -(f+1)$, respectively, and we obtain (recalling $J_{-n} = (-1)^n J_n$):

$$\frac{d\gamma}{dt} = \frac{e\hat{K}|\hat{E}|}{\gamma mc}\mathcal{R}e\left\{e^{i(f\xi+\phi)}\left[J_{-\frac{f-1}{2}}(f\eta) + J_{-\frac{f+1}{2}}(f\eta)\right]\right\} \tag{6.12}$$

or $$\frac{d\gamma}{dt} = \frac{e\hat{K}|\hat{E}|}{\gamma mc}\cos(f\xi+\phi)\cdot\left[J_{\frac{f-1}{2}}(f\eta) - J_{\frac{f+1}{2}}(f\eta)\right](-1)^{\frac{f-1}{2}}. \tag{6.13}$$

This equation occupies the most general starting point in FEL analysis and is used directly in the analysis of tapered or inverse-tapered FELs (Brau 1990, Colson 1981) for which $\Delta\gamma$ can be a substantial fraction of γ in a single pass through the undulator. The corresponding electron phase space coordinates are (ξ, γ).

6.2 The FEL pendulum equation

However, for uniform undulators, changes in γ on the lhs of (6.13) are generally much smaller than the overall γ appearing in the denominator on the rhs (e.g. see section 11.5, (11.43)) and the FEL interaction remains essentially resonant except for tiny changes in γ that correspond to longitudinal drifts in (ξ, ν) phase space. Therefore, we treat the γ appearing on the rhs as a constant, specifically, the incident γ.

To treat the lhs of (6.13), note from the definition of ν in (6.8) that if γ is constant, then $\bar{\beta}_z$ is constant and ν is constant. In general, changes in γ are related to changes in ν. We thus seek to relate $d\gamma/dt$ on the lhs to the derivative $d\nu/d\tau$ of the phase velocity. Recall from (4.15) that

$$\bar{\beta}_z = 1 - \frac{1+\hat{K}^2}{2\gamma^2} \Rightarrow \frac{d\bar{\beta}_z}{dt} = \frac{1+\hat{K}^2}{2\gamma^2}\left(\frac{2}{\gamma}\right)\frac{d\gamma}{dt} \Rightarrow \frac{d\bar{\beta}_z}{dt} = \frac{2k_w}{\gamma k}\frac{d\gamma}{dt}, \tag{6.14}$$

where we used (4.16) to insert $\lambda/\lambda_w = k_w/k$. This gives a relation between $\mathrm{d}\bar{\beta}_z/\mathrm{d}t$ and $\mathrm{d}\gamma/\mathrm{d}t$. We can also independently relate $\mathrm{d}\bar{\beta}_z/\mathrm{d}t$ to $\mathrm{d}\nu/\mathrm{d}\tau$ as follows,

$$\frac{\mathrm{d}\nu}{\mathrm{d}t}\left(=\frac{c}{L_w}\frac{\mathrm{d}\nu}{\mathrm{d}\tau}\right) = L_w(k+k_w)\frac{\mathrm{d}\bar{\beta}_z}{\mathrm{d}t} = L_w k\frac{\mathrm{d}\bar{\beta}_z}{\mathrm{d}t}, \tag{6.15}$$

where we dropped $k_w \ll k$ in the last equality. Combining (6.14) and (6.15), we thus obtain

$$\frac{\mathrm{d}\nu}{\mathrm{d}t} = \frac{2L_w k_w}{\gamma}\frac{\mathrm{d}\gamma}{\mathrm{d}t}, \qquad \text{or} \qquad \frac{\mathrm{d}\gamma}{\mathrm{d}t} = \frac{\gamma c}{2k_w L_w^2}\frac{\mathrm{d}\nu}{\mathrm{d}\tau}, \tag{6.16}$$

which gives the desired relation. Substituting this expression for $\mathrm{d}\gamma/\mathrm{d}t$ into (6.13), we obtain

$$\frac{\mathrm{d}\nu}{\mathrm{d}\tau} = \frac{e\hat{K}|\hat{E}|2k_w L_w^2}{\gamma^2 mc^2}\cos(f\xi+\phi)\cdot\left[J_{\frac{f-1}{2}}(f\eta) - J_{\frac{f+1}{2}}(f\eta)\right](-1)^{\frac{f-1}{2}} \tag{6.17}$$

$$= \frac{4\pi e N_w^2\lambda_w\hat{K}}{\gamma^2 mc^2}|\hat{E}|\cos(f\xi+\phi)\cdot\left[J_{\frac{f-1}{2}}(f\eta) - J_{\frac{f+1}{2}}(f\eta)\right](-1)^{\frac{f-1}{2}}, \tag{6.18}$$

where we inserted $k_w = 2\pi/\lambda_w$ and $L_w = N_w\lambda_w$ in the second line. Finally, we introduce the *dimensionless optical field* 'a',

$$a \equiv |a|\mathrm{e}^{\mathrm{i}\phi} = \frac{4\pi e N_w^2\lambda_w\hat{K}_f}{\gamma^2 mc^2}|\hat{E}|\mathrm{e}^{\mathrm{i}\phi}, \tag{6.19}$$

where

$$\hat{K}_f \equiv \hat{K}\cdot\left[J_{\frac{f-1}{2}}(f\eta) - J_{\frac{f+1}{2}}(f\eta)\right](-1)^{\frac{f-1}{2}} \tag{6.20}$$

$$\left(\hat{K}_f = \hat{K}\cdot\left[J_0(\eta) - J_1(\eta)\right] \quad \text{for } f=1.\right) \tag{6.21}$$

The equation describing the electron phase space evolution in the presence of the optical field is therefore

$$\frac{\mathrm{d}\nu}{\mathrm{d}\tau} = \frac{\mathrm{d}^2\xi}{\mathrm{d}\tau^2} = |a|\cos(f\xi+\phi). \tag{6.22}$$

The equation is called the dimensionless *FEL pendulum equation*, and is one of the fundamental equations in FEL physics. Its validity spans all magnitudes $|a|$ of the dimensionless optical field for which $\Delta\gamma \ll \gamma$.

References

Brau C A 1990 *Free-Electron Lasers* (Boston, MA: Academic)

Colson W B 1981 The nonlinear wave equation for higher harmonics in free-electron lasers *IEEE J. Quantum Electron.* **QE-17** 1417–27

Chapter 7

Effect of electron motion on the optical field

7.1 The wave equation

The derivation of the FEL wave equation in sections 7.1–7.3 is again based on the analysis of Colson (1981). In this analysis, the evolution and amplification of the co-propagating optical wave are obtained directly from the wave equation. The most convenient gauge is the Coulomb gauge. In this gauge the source term for the vector potential $\vec{A}$ differs from the physical current density $\vec{J}$ by a term involving the scalar potential Φ (Jackson 1975). Since the Coulomb field was omitted in the analysis of electron motion on physical grounds (sections 4.1 and 6.1), the term involving Φ must be self-consistently omitted from the wave equation. The wave equation is thus driven only by the physical current density $\vec{J}$, consisting of a transverse component $\vec{J}_\perp$, and we write

$$\left(\nabla^2 - \frac{1}{c^2}\frac{\partial^2}{\partial t^2}\right)\vec{A} = -\frac{4\pi}{c}\vec{J}_\perp, \tag{7.1}$$

where the optical electric field $\vec{E}_r(t)$ is obtained from the optical vector potential $\vec{A}\,(t)$ by

$$\vec{E}_r = -\frac{1}{c}\frac{\partial \vec{A}}{\partial t}. \tag{7.2}$$

To simplify the subsequent analysis, we introduce the following form of the vector potential in cylindrical coordinates $(\vec{r}, z, t)$,

$$\vec{A} = \vec{A}(\vec{r}, z, t) = \hat{x}\frac{|E(\vec{r}, z, t)|}{fk}\sin\big(fkz - f\omega t + \phi(\vec{r}, z, t)\big), \tag{7.3}$$

where the optical wavenumber $k = \omega/c$, and $E = |E|e^{i\phi}$. Differentiation of this expression by $-\frac{1}{c}\frac{\partial}{\partial t}$ returns to us the original form of the co-propagating optical field that we assumed in the previous analyses,

$$\vec{E}_r = \hat{x}|E(\vec{r}, z, t)|\cos\big(fkz - f\omega t + \phi(\vec{r}, z, t)\big) + \text{slowly varying terms in } \frac{\partial|E|}{\partial t}, \frac{\partial\phi}{\partial t}, \tag{7.4}$$

if we neglect the slowly varying terms. We can in fact make this approximation in essentially all physical systems of interest and it is the same approximation we will employ below, so we remain consistent. The approximation is called the *slowly varying envelope approximation* (SVEA), and for a complex envelope function $E(t) = |E(t)|e^{i\phi(t)}$ it allows us to neglect terms for which

$$\frac{\partial E}{\partial t} \ll \omega E \tag{7.5}$$

$$\frac{\partial^2 E}{\partial t^2} \ll \omega\frac{\partial E}{\partial t}. \tag{7.6}$$

Physically, it states that the envelope changes significantly only on time scales much longer than the period of the carrier wave, or equivalently, that the fractional change in the envelope over one wavelength satisfies $\Delta|E|/|E| \ll 1$; $\Delta\phi \ll 2\pi$, as depicted in figure 7.1.

Let's proceed with the wave equation; write (7.3) in terms of the complex envelope E as

$$\vec{A} = \hat{x}A_x = \hat{x}\frac{E}{2ifk}e^{i(fkz - f\omega t)} + \text{c.c.}, \qquad \text{where } E = |E|e^{i\phi}. \tag{7.7}$$

Then the derivatives are:

$$\left(\frac{\partial^2}{\partial x^2} + \frac{\partial^2}{\partial y^2}\right)A_x \equiv \nabla_\perp^2 A_x = \frac{e^{i(fkz - f\omega t)}}{2ifk}\nabla_\perp^2 E + \text{c.c.} \tag{7.8}$$

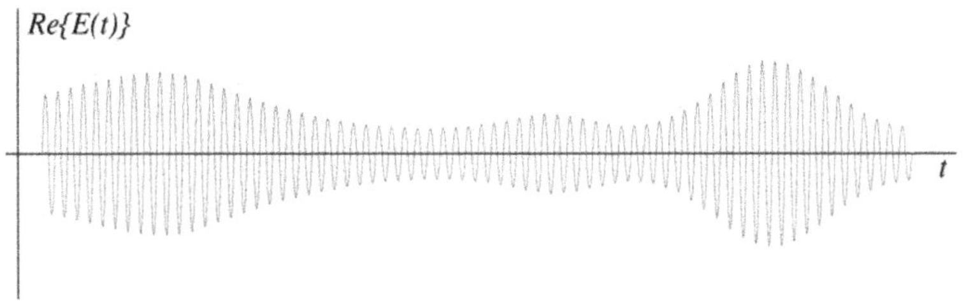

Figure 7.1. Depiction of a wave satisfying the SVEA.

$$\frac{\partial^2 A_x}{\partial z^2} = \mathrm{i}fk\frac{E}{2}\mathrm{e}^{\mathrm{i}(fkz-f\omega t)} + \frac{\partial E}{\partial z}\mathrm{e}^{\mathrm{i}(fkz-f\omega t)} + \frac{1}{2\mathrm{i}fk}\frac{\partial^2 E}{\partial z^2}\mathrm{e}^{\mathrm{i}(fkz-f\omega t)} + \text{c.c.} \tag{7.9}$$

$$\frac{\partial^2 A_x}{\partial t^2} = \mathrm{i}f\frac{\omega^2}{k}\frac{E}{2}\mathrm{e}^{\mathrm{i}(fkz-f\omega t)} - \frac{\omega}{k}\frac{\partial E}{\partial t}\mathrm{e}^{\mathrm{i}(fkz-f\omega t)} + \frac{1}{2\mathrm{i}fk}\frac{\partial^2 E}{\partial t^2}\mathrm{e}^{\mathrm{i}(fkz-f\omega t)} + \text{c.c.} \tag{7.10}$$

The first term in each of the latter expressions cancels against the other in the wave equation (using $k = \omega/c$), and the wave equation (7.1) reduces to

$$\hat{x}\left[\frac{1}{2\mathrm{i}fk}\left(\nabla_\perp^2 E + \frac{\partial^2 E}{\partial z^2}\right) + \left(\frac{\partial E}{\partial z} + \frac{1}{c}\frac{\partial E}{\partial t}\right) - \frac{1}{c}\frac{1}{2\mathrm{i}f\omega}\frac{\partial^2 E}{\partial t^2}\right]\mathrm{e}^{\mathrm{i}(fkz-f\omega t)} + \text{c.c.} = -\frac{4\pi}{c}\vec{J}_\perp$$

$$\begin{array}{ccccc} \uparrow & \uparrow & \uparrow & \uparrow & \uparrow \\ A & B & C & D & E \leftarrow \text{examine these terms} \end{array} \tag{7.11}$$

We may neglect E with respect to D by the SVEA; we must retain both C and D to complete the description of the longitudinal evolution; but as far as the transverse evolution is concerned, we may drop B with respect to A.

Neglecting $\partial^2/\partial z^2$ with respect to $\partial^2/\partial x^2$ and $\partial^2/\partial y^2$ is another physical approximation, different from the SVEA, that is appropriate for describing a beam that diverges or converges slowly in space; it is called the *paraxial wave approximation* and applies in essentially all physical situations of interest (figure 7.2).

(If we were considering only the unphysical case of plane waves, for which $\nabla_\perp^2 = 0$, we would then drop $(1/k)\partial^2/\partial z^2$ with respect to $\partial/\partial z$ by the SVEA.) The wave equation (7.11) becomes

$$\hat{x}\left[\frac{1}{2\mathrm{i}fk}\nabla_\perp^2 E + \left(\frac{\partial E}{\partial z} + \frac{1}{c}\frac{\partial E}{\partial t}\right)\right] + \hat{x}[\ldots]^*\mathrm{e}^{-2\mathrm{i}(fkz-f\omega t)} = -\frac{4\pi}{c}\vec{J}_\perp\,\mathrm{e}^{-\mathrm{i}(fkz-f\omega t)}, \tag{7.12}$$

where we multiplied both sides by the complex conjugate of the exponential factor in the first term.

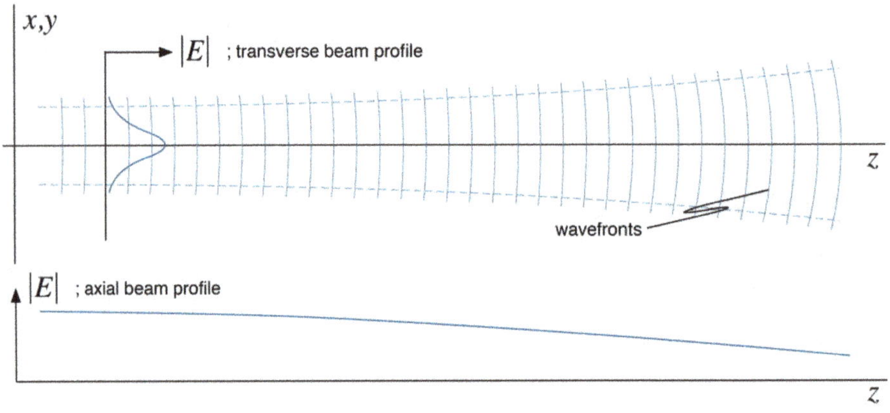

Figure 7.2. Depiction of a beam satisfying the paraxial wave approximation.

7.2 Transverse currents

Consider a single electron, the ith electron, for which (recalling (4.8))

$$\vec{J}_\perp \equiv -e\vec{v}_\perp \delta^{(3)}(\vec{r} - \vec{r}_i) = +\hat{x}\,\frac{ecK}{\gamma}\cos(k_w z)\delta^{(3)}(\vec{r} - \vec{r}_i) \tag{7.13}$$

$$= +\hat{x}\,\frac{ecK}{2\gamma_i}\left[\mathrm{e}^{+\mathrm{i}k_w z_i} + \mathrm{e}^{-\mathrm{i}k_w z_i}\right]\delta^{(3)}(\vec{r} - \vec{r}_i), \tag{7.14}$$

where we used the sifting property of the delta function, $f(x)\delta(x-a) \equiv f(a)\delta(x-a)$, in the second line. Then the $\hat{x}$ component of the wave equation, driven by this single electron at position z_i, is

$$\left[\frac{1}{2\mathrm{i}fk}\nabla_\perp^2 E + \left(\frac{\partial E}{\partial z} + \frac{1}{c}\frac{\partial E}{\partial t}\right)\right] + [\ldots]^* \mathrm{e}^{-2\mathrm{i}(fkz - f\omega t)}$$
$$= -\frac{2\pi eK}{\gamma_i}\delta^{(3)}(\vec{r} - \vec{r}_i)\left[\mathrm{e}^{\mathrm{i}k_w z_i}\mathrm{e}^{-\mathrm{i}(fkz_i - f\omega t)} + \mathrm{e}^{-\mathrm{i}k_w z_i}\mathrm{e}^{-\mathrm{i}(fkz_i - f\omega t)}\right]. \tag{7.15}$$

This equation involves fast oscillatory terms that do not contribute to the mean growth of the wave along the undulator. We therefore again perform an average over one magnet period to eliminate these fast oscillations as we did in section 6.1 for the electron evolution. When this is done on the lhs of the above equation, the conjugated term vanishes and on the rhs the Bessel terms appear in precisely the same manner as in section 6.1, the algebra being identical. We obtain

$$\frac{1}{2\mathrm{i}fk}\nabla_\perp^2 \hat{E} + \left(\frac{\partial \hat{E}}{\partial z} + \frac{1}{c}\frac{\partial \hat{E}}{\partial t}\right) = -\frac{2\pi e\hat{K}_f}{\gamma_i}\delta^{(3)}(\vec{r} - \vec{r}_i)\mathrm{e}^{-\mathrm{i}f\xi_i};\ \xi_i \equiv (k_w + k)\bar{z}_i - \omega t, \tag{7.16}$$

where, as before,

$$\hat{K}_f \equiv \hat{K}\cdot(-1)^{\frac{f-1}{2}}\left[J_{\frac{f-1}{2}}(f\eta) - J_{\frac{f+1}{2}}(f\eta)\right], \tag{7.17}$$

and we have introduced rms values $(\hat{E}, \hat{K})$ on each side of the equation.

We now include the total contribution to the transverse current from all electrons i within a small volume ΔV with dimensions on the order of an optical wavelength. Since the wave equation is linear, these currents are formally added on the rhs of (7.16) and we obtain

$$\frac{1}{2\mathrm{i}fk}\nabla_\perp^2 \hat{E} + \left(\frac{\partial \hat{E}}{\partial z} + \frac{1}{c}\frac{\partial \hat{E}}{\partial t}\right) = -2\pi e\hat{K}_f \sum_i \delta^{(3)}(\vec{r} - \vec{r}_i)\frac{\mathrm{e}^{-\mathrm{i}f\xi_i}}{\gamma_i}. \tag{7.18}$$

Recall that the average of any quantity b over the discrete electron density within this volume can be written

$$\langle b\rangle_{\Delta V} \equiv \frac{\sum_i b_i\,\delta^{(3)}(\vec{r} - \vec{r}_i)}{\sum_i \delta^{(3)}(\vec{r} - \vec{r}_i)} \equiv \frac{\sum_i b_i\,\delta^{(3)}(\vec{r} - \vec{r}_i)}{n_e} \tag{7.19}$$

by definition of the electron volume density $n_e = \sum_i \delta^{(3)}(\vec{r} - \vec{r}_i)$. Thus, with $b_i = \mathrm{e}^{-\mathrm{i}f\xi_i}/\gamma_i$, the wave equation becomes

$$\frac{1}{2\mathrm{i}fk}\nabla_\perp^2 \hat{E} + \left(\frac{\partial \hat{E}}{\partial z} + \frac{1}{c}\frac{\partial \hat{E}}{\partial t}\right) = -2\pi e \hat{K}_f\, n_e \left\langle \frac{\mathrm{e}^{-\mathrm{i}f\xi}}{\gamma} \right\rangle_{\Delta V}, \tag{7.20}$$

where the average is over the electrons i within a small volume ΔV with dimensions on the order of an optical wavelength. Since each electron is uniquely identified by its initial position (ξ_0, ν_0) in phase space, this is formally equivalent to taking an average over the initial phase space distribution, $\langle \ldots \rangle_{\Delta V} \rightarrow \langle \ldots \rangle_{\xi_0, \nu_0}$. Together with (6.13), (7.20) occupies the most general starting point in FEL analysis and is employed in systems for which $\Delta\gamma$ can be a substantial fraction of γ.

7.3 The FEL wave equation

To describe the FEL interaction for Gaussian beams of finite spatial extent, the full wave equation (7.20) must be employed and we consider this in chapter 8. However, much of basic FEL physics can be apprehended by considering only the longitudinal dependence of the interaction. For this case, we consider plane waves for which the transverse Laplacian $\nabla_\perp^2 = 0$. The full wave equation then reduces to

$$\frac{\partial \hat{E}}{\partial z} + \frac{1}{c}\frac{\partial \hat{E}}{\partial t} = -2\pi e \hat{K}_f\, n_e \left\langle \frac{\mathrm{e}^{-\mathrm{i}f\xi}}{\gamma} \right\rangle_{\xi_0, \nu_0}. \tag{7.21}$$

At this point it is convenient to make a coordinate transformation in the laboratory frame to 'co-moving' axes that travel along with the optical beam (figure 7.3). We consider a volume ΔV of the beam which is small compared with the spatial or temporal variations in either the optical- and electron-beam envelopes.

The formal coordinate transformation is

$$\tilde{z} = z - ct \tag{7.22}$$

$$\tilde{x} = x \tag{7.23}$$

$$\tilde{y} = y \tag{7.24}$$

$$\tilde{t} = t. \tag{7.25}$$

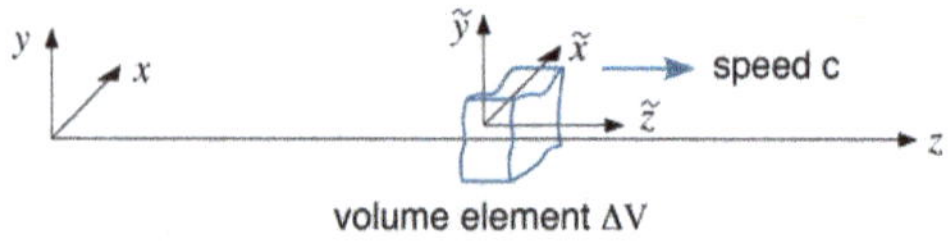

Figure 7.3. Co-moving coordinates in the laboratory frame.

Note that the transformed coordinate $\tilde{z}$ parametrizes the 'microscopic' position within the beam, while the coordinate $\tilde{t}$, formally equal to t, parametrizes the distance along the undulator.

This transformation is not a Lorentz transformation; mathematically, it is Galilean. Thus, it does not describe what a co-moving observer would see; that is not what it is supposed to do. We are still describing the interaction in the laboratory frame, but merely with respect to a more convenient coordinate system—one that happens to be moving. Thus, for example, a given optical pulse has the same duration in both coordinate systems (say, 2 ps), and the same optical intensity, etc. Formally, the only effect of this transformation is to convert

$$\frac{\partial E}{\partial z} + \frac{1}{c}\frac{\partial E}{\partial t} \rightarrow \frac{\partial \tilde{E}}{\partial \tilde{t}}, \tag{7.26}$$

where $E(x, y, z, t) = \tilde{E}(\tilde{x}, \tilde{y}, \tilde{z}, \tilde{t})$, and *that* is its only purpose. With the tildes suppressed, the wave equation for plane waves reduces to

$$\frac{1}{c}\frac{\mathrm{d}\hat{E}}{\mathrm{d}t} = -2\pi e\hat{K}_f\, n_e \left\langle \frac{\mathrm{e}^{-\mathrm{i}\xi}}{\gamma} \right\rangle_{\xi_0,\nu_0}, \tag{7.27}$$

where the total derivative reflects dependence on time only.

Co-moving coordinates are useful because they allow us to describe the FEL interaction between a particular section of the optical wave and the associated sample of co-propagating electrons. But how do we account for slippage? If we have CW electron and optical beams in which end-effects are absent, and if microtemporal perturbations such as electron shot noise or the sideband instability (see section 11.2) are neglected, then slippage does not need to be considered separately: the co-moving coordinate t can be equivalently used to track both the optical wave and the interacting electrons as they travel together along the undulator. In essentially all systems of interest, the optical envelope in any beam section of length $N_w\lambda$ will be uniform regardless of slippage and electrons in adjacent beam sections of length $N_w\lambda$ will be driven by the same field $\hat{E}(t)$. This uniformity arises from the fact that the slippage length $N_w\lambda$ is much smaller than the distance required for the fields and electrons to appreciably evolve, which is on the order of the magnet period λ_w. However, when we describe the interaction of short pulses in chapter 15 in which various microtemporal and end-effects are manifest, the coupled equations describing the evolution of the electrons and optical wave will have to be separately re-written to include slippage explicitly.

If we now assume $\gamma \approx$ constant on a single pass, we can introduce into (7.27) the dimensionless quantities

$$\tau = \frac{ct}{L_w}; \qquad a = |a|\mathrm{e}^{\mathrm{i}\phi} = \frac{4\pi e N_w^2 \lambda_w \hat{K}_f}{\gamma^2 mc^2}|\hat{E}|\mathrm{e}^{\mathrm{i}\phi}; \qquad j = \frac{8\pi^2 e^2 N_w^3 \lambda_w^2 \hat{K}_f^2}{\gamma^3 mc^2} n_e, \tag{7.28}$$

where τ is the dimensionless time (parameterizing the distance along the undulator), a is the dimensionless optical field from (6.19) and j is the *dimensionless current density*. We then obtain the dimensionless *FEL wave equation*

$$\frac{da}{d\tau} = -j\langle e^{-if\xi}\rangle_{\xi_0,\nu_0}. \tag{7.29}$$

This equation and the FEL pendulum equation derived in chapter 6 together comprise the *coupled Maxwell–Lorentz equations of motion.* These equations self-consistently describe the classical fundamental FEL interaction and often serve as the starting point for many calculations in the field of FELs.

We should make one more point here regarding the evolution of the optical phase $\phi(\tau)$. The optical field $a(\tau)$ is of course a complex quantity and the wave equation, (7.29), includes the slow evolution of $\phi(\tau)$ along the undulator. But the CW equations as written do not otherwise incorporate microtemporal variations in the phase $\phi(\tilde{z}, \tau)$ as a function of the co-moving coordinate $\tilde{z}$. These latter variations are the manifestation of an actual variation in the optical frequency ω as the FEL interaction evolves in τ, where $\omega(\tau) = c\partial\phi(\tilde{z}, \tau)/\partial\tilde{z}$. This phenomenon includes both the evolution of the optical frequency towards the peak of the gain curve in the small signal regime and the phenomenon known as 'frequency pulling' at saturation (see section 11.2). To properly track this evolution of the optical frequency in simulations, slippage must again be explicitly included in the equations of motion.

7.4 Energy conservation

The FEL pendulum equation describes how the electrons evolve in (ξ, ν)-space in the presence of the co-propagating optical field $a = |a|e^{i\phi}$. The FEL wave equation describes how the optical field is driven by the resulting evolution of the co-propagating electrons in phase space. The self-consistency of these coupled equations of motion is illustrated by energy conservation (Elleaume and Deacon 1984). Consider the plane-wave equations

$$\frac{d\nu}{d\tau} = |a|\cos(f\xi + \phi) \tag{7.30}$$

$$\frac{da}{d\tau} = -j\langle e^{-if\xi}\rangle_{\xi_0,\nu_0}, \tag{7.31}$$

which are valid at any time τ, for any field a and for any current density j. The initial phase space coordinates (ξ_0, ν_0) are those of the electrons within a sample volume ΔV with dimensions on the order of an optical wavelength. From the pendulum equation (7.30) write

$$\left\langle\frac{d\nu}{d\tau}\right\rangle_{\xi_0,\nu_0} = \left\langle|a|\frac{1}{2}e^{if\xi}e^{i\phi} + |a|\frac{1}{2}e^{-if\xi}e^{-i\phi}\right\rangle_{\xi_0,\nu_0} \tag{7.32}$$

$$= \frac{a}{2}\langle e^{if\xi}\rangle_{\xi_0,\nu_0} + \frac{a^*}{2}\langle e^{-if\xi}\rangle_{\xi_0,\nu_0}, \tag{7.33}$$

where we used $a = |a|e^{i\phi}$. Multiply by $-2j$ and substitute for the averaged exponentials on the rhs from the wave equation (7.31):

$$-2j\left\langle \frac{d\nu}{d\tau} \right\rangle_{\xi_0,\nu_0} = a\frac{da^*}{d\tau} + a^*\frac{da}{d\tau} = \frac{d|a|^2}{d\tau} \tag{7.34}$$

$$\text{or} \quad -2j\,\langle d\nu \rangle_{\xi_0,\nu_0} = d|a|^2. \tag{7.35}$$

This is energy conservation, equating the energy lost by the electrons in the sample volume ΔV to the energy gained by the optical wave in the same volume. To see this more explicitly, note from the relation between $d\gamma$ and $d\nu$, (6.16), that

$$d\nu = 2L_w k_w \frac{d\gamma}{\gamma} = 4\pi N_w \frac{d\gamma}{\gamma}, \tag{7.36}$$

and consider the N_e electrons in volume ΔV that contribute to the average $\langle ... \rangle_{\Delta V} \equiv \langle ... \rangle_{\xi_0,\nu_0}$. Inserting (7.36) and the explicit expressions for a and j from (7.28) into (7.35), we obtain

$$-2\frac{8\pi^2 e^2 N_w^3 \lambda_w^2 \hat{K}_f^2}{\gamma^3 mc^2} n_e\, 4\pi N_w \left\langle \frac{d\gamma}{\gamma} \right\rangle_{\Delta V} = \frac{16\pi^2 e^2 N_w^4 \lambda_w^2 \hat{K}_f^2}{\gamma^4 m^2 c^4} d|\hat{E}|^2. \tag{7.37}$$

Almost everything cancels, except for

$$-n_e\, mc^2 \langle d\gamma \rangle_{\Delta V} = \frac{1}{4\pi} d|\hat{E}|^2, \tag{7.38}$$

$$\text{or, writing } n_e \equiv \frac{N_e}{\Delta V}: \quad -N_e \langle d(\gamma mc^2) \rangle_{\Delta V} = d\left(\frac{1}{4\pi} |\hat{E}|^2 \Delta V \right). \quad \text{QED} \tag{7.39}$$

This calculation of energy conservation will be re-visited in sections 8.3 and 8.4 when the equations of motion are extended to include the transverse dependence of the optical and electron beams. Energy conservation will be also employed directly in the calculation of small-signal gain in chapter 9 and in the calculation of FEL extraction efficiency and optical power at saturation in chapters 11 and 12.

References

Colson W B 1981 The nonlinear wave equation for higher harmonics in free-electron lasers *IEEE J. Quantum Electron.* **QE-17** 1417–27

Elleaume P and Deacon D A G 1984 Transverse mode dynamics in a free-electron laser *Appl. Phys.* B **33** 9–16

Jackson J D 1975 *Classical Electrodynamics* 2nd edn (New York: Wiley)

Chapter 8

Transverse modes in the equations of motion

8.1 Superposition of transverse modes

The transverse mode analysis in sections 8.1–8.3 is based on the paper of Elleaume and Deacon (1984). We start with the wave equation including the full transverse dependence of the optical wave, (7.20), in the form

$$\frac{1}{2\mathrm{i}fk}\nabla_{\perp}^{2}\hat{E}+\left(\frac{\partial\hat{E}}{\partial z}+\frac{1}{c}\frac{\partial\hat{E}}{\partial t}\right)=-\frac{2\pi e\hat{K}_{f}}{\gamma}n_{e}\left\langle \mathrm{e}^{-\mathrm{i}f\xi}\right\rangle_{\xi_{0},\nu_{0}},\tag{8.1}$$

where we average over initial phase space coordinates (ξ_0, ν_0) and assume $\Delta\gamma \ll \gamma$. The solutions to the stationary ($\frac{\partial\hat{E}}{\partial t} = 0$) and homogeneous (rhs $=0$) form of this equation comprise a set of transverse eigenmodes that satisfy the boundary conditions of the physical system of interest, say, an optical resonator or some other structure. In many cases, these eigenmodes can be closely approximated by one of a complete basis set of orthogonal transverse modes, say, the free-space Hermite–Gauss or Gauss–Laguerre modes; more generally, the eigenmodes of the system are composed of a superposition of the basis modes. Siegman (1986) contains an extensive discussion of the form and properties of the free-space transverse modes and their application to various spatial resonator structures and symmetries; in our analysis, we consider the cylindrically symmetric Gauss–Laguerre modes, which are the ones usually encountered in FEL physics.

The entire basis set of orthonormal modes is uniquely specified, for the assumed frequency $f\omega$ or wavenumber $fk = f\omega/c$ appearing in the wave equation, by the *location of the waist* and the *Rayleigh range* z_R (related to the beam radius at the waist). Any choice of these two parameters yields a complete basis set of modes appropriate for the superposition of any particular solution; however, the number of basis modes needed to represent a given solution is minimized by a judicious choice of these parameters. In essentially all cases, the appropriate choice is the one for which the wavefront radius of curvature of the basis modes matches the radius of

doi:10.1088/978-1-6270-5573-4ch8

curvature of the optical resonator mirrors. A given eigenmode of the system is then dominated by a particular orthonormal mode of the basis set. FELs having long, narrow undulator vacuum bores effectively spatially filter out all but the lowest order eigenmode, which is dominated by the TEM_{00} basis mode.

The individual modes in the basis set are specified by a pair of mode indices $\{p, m\}$, p for radial and m for azimuthal. We consider only the radially symmetric Gauss–Laguerre modes in our analyses, for which $m = 0$.

With the azimuthal index m suppressed, these modes have the following form in cylindrical coordinates:

$$u_p(r, z) = E_p(r, z)\mathrm{e}^{\mathrm{i}\psi_p(r,z)}; \quad p = 0, 1, 2, \ldots \tag{8.2}$$

with
$$E_p(r, z) = \sqrt{\frac{2}{1+\zeta^2}} \cdot L_p\left(\frac{2\rho^2}{1+\zeta^2}\right) \cdot \exp\left[-\frac{\rho^2}{1+\zeta^2}\right]; \quad \rho^2 \equiv \frac{r^2}{w_0^2} = \frac{x^2+y^2}{w_0^2} \tag{8.3}$$

$$\psi_p(r, z) = \frac{\rho^2 \zeta}{1+\zeta^2} - (2p+1)\tan^{-1}\zeta; \qquad \zeta \equiv \frac{z}{z_R}; \qquad z_R = \frac{\pi w_0^2}{\lambda_f}, \tag{8.4}$$

where L_p is the pth Laguerre polynomial and the distance z is measured from the waist of the beam (where the beam radius is w_0). These modes are solutions of the homogeneous wave equation at the harmonic wavelength $\lambda_f = \lambda/f$,

$$\nabla_\perp^2 u_p(r, z) + 2\mathrm{i}fk\frac{\partial u_p(r, z)}{\partial z} = 0; \qquad fk = \frac{f\omega}{c}, \tag{8.5}$$

and satisfy the orthonormality condition

$$\iint_{-\infty}^{\infty} \frac{\mathrm{d}x\,\mathrm{d}y}{\pi w_0^2} u_q^*(x, y; z)u_p(x, y; z) = \int_0^{\infty} \mathrm{d}(\rho^2)u_q^*(\rho, z)u_p(\rho, z) = \delta_{pq}. \tag{8.6}$$

They are the complex conjugates of the modes defined by Siegman (1986), because his choice of phase for the optical wave is $\exp \mathrm{i}(\omega t - kz + \phi)$, while our choice in this text is $\exp \mathrm{i}(kz - \omega t + \phi)$; see (6.2).

To solve the full wave equation, (8.1), expand the solution in transverse modes u_p as follows:

$$\hat{E}(r, z, t) = |\hat{E}(r, z, t)|\mathrm{e}^{\mathrm{i}\phi(r,z,t)} \equiv \sum_p c_p(t)u_p(r, z), \text{ with } c_p(t) = |c_p(t)|\mathrm{e}^{\mathrm{i}\phi_p(t)}, \tag{8.7}$$

where the explicit time dependence is contained in the expansion coefficients $c_p(t)$, which do not contain any transverse dependence. The solution of the wave equation will consist of solving for the time-dependence of the expansion coefficients $c_p(t)$.

It is useful and important to keep in mind the meaning of the time dependence of the coefficients $c_p(t)$. Say we have an 'electron beam' consisting of a single electron, which has no effect on the co-propagating optical beam. The optical field and intensity at the position of the electron will certainly change with time along the undulator as the beam is focused through the waist. However, in the absence of any spatial perturbations such as apertures, etc, the mode decomposition *will not change*: if we compute the expansion coefficients at two different times t_1 and t_2, the two sets of coefficients will be identical, even though the basis functions will have changed due to their dependence on z.

Now consider a full, co-propagating electron beam: we compute the expansion coefficients at time t_1 and then we let the electron and optical beams interact along the undulator until time t_2. The optical power will have increased due to the exchange of energy with the electrons; this increase must be reflected in the mode coefficients $c_p(t)$, since the basis modes are normalized by definition. Moreover, we suspect that the optical beam will be transversely distorted by the electron beam, because, for example, if the electron beam is much narrower than the optical beam and exhibits high gain, then the optical beam will be non-uniformly amplified near the axis compared to the fringes. A projection onto the basis modes will thus yield a different set of expansion coefficients, $\frac{c_p}{c_q}|_{t_1} \neq \frac{c_p}{c_q}|_{t_2}$. The entire set of expansion coefficients will therefore change with time; this is the origin of their time dependence.

8.2 The mode evolution equation

Substitute the mode expansion for $\hat{E}$ from (8.7) into the wave equation (8.1):

$$\frac{1}{2\mathrm{i}fk}\nabla_\perp^2\sum_p c_p u_p + \left(\frac{\partial}{\partial z}\sum_p c_p u_p + \frac{1}{c}\frac{\partial}{\partial t}\sum_p c_p u_p\right) = -\frac{2\pi e\hat{K}_f}{\gamma} n_e \langle \mathrm{e}^{-\mathrm{i}f\xi}\rangle_{\xi_0,\nu_0} \tag{8.8}$$

$$\frac{1}{2\mathrm{i}fk}\sum_p c_p \nabla_\perp^2 u_p + \sum_p c_p \frac{\partial u_p}{\partial z} + \frac{1}{c}\sum_p \frac{\mathrm{d}c_p}{\mathrm{d}t} u_p = -\frac{2\pi e\hat{K}_f}{\gamma} n_e \langle \mathrm{e}^{-\mathrm{i}f\xi}\rangle_{\xi_0,\nu_0}. \tag{8.9}$$

Now rearrange and collect terms:

$$\sum_p c_p\left[\frac{1}{2\mathrm{i}fk}\nabla_\perp^2 u_p + \frac{\partial u_p}{\partial z}\right] + \frac{1}{c}\sum_p \frac{\mathrm{d}c_p}{\mathrm{d}t} u_p = -\frac{2\pi e\hat{K}_f}{\gamma} n_e \langle \mathrm{e}^{-\mathrm{i}f\xi}\rangle_{\xi_0,\nu_0}. \tag{8.10}$$

The first summation drops out because the u_p are solutions to the homogeneous wave equation, (8.5). Projecting the remaining summation onto the qth mode yields

$$\frac{1}{c}\sum_p \frac{\mathrm{d}c_p}{\mathrm{d}t}\iint_{-\infty}^{\infty}\frac{\mathrm{d}x\,\mathrm{d}y}{\pi w_0^2}u_q^* u_p = -\frac{2\pi e\hat{K}_f}{\gamma}\iint_{-\infty}^{\infty}\frac{\mathrm{d}x\,\mathrm{d}y}{\pi w_0^2}u_q^*\, n_e(x, y, t)\langle \mathrm{e}^{-\mathrm{i}f\xi}\rangle_{\xi_0,\nu_0} \tag{8.11}$$

$$\frac{1}{c}\frac{\mathrm{d}c_q}{\mathrm{d}t} = -\frac{2\pi e\hat{K}_f}{\gamma}\iint_{-\infty}^{\infty}\frac{\mathrm{d}x\,\mathrm{d}y}{\pi w_0^2}u_q^*\, n_e(x, y, t)\langle \mathrm{e}^{-\mathrm{i}f\xi}\rangle_{\xi_0,\nu_0}, \tag{8.12}$$

where the orthonormality of the modes, (8.6), reduced the summation on the lhs of (8.11) to a single term. Separate the electron density n_e into longitudinal and transverse factors

$$n_e(x, y, t) \equiv n_L(t) n_{\mathrm{T}}(x, y, t), \tag{8.13}$$

with all of the transverse dependence contained in the factor n_{T}, and introduce the dimensionless parameters

$$a_p(\tau) = \frac{4\pi e N_w^2 \lambda_w \hat{K}_f}{\gamma^2 mc^2} c_p(\tau); \qquad a_p = |a_p| \mathrm{e}^{\mathrm{i}\phi_p} \tag{8.14}$$

$$j_L(\tau) = \frac{8\pi^2 e^2 N_w^3 \lambda_w^2 \hat{K}_f^2}{\gamma^3 mc^2} n_L(\tau), \tag{8.15}$$

where $\tau = ct/L_w$. Then the expansion coefficients satisfy the *mode evolution equation*

$$\frac{\mathrm{d}a_p}{\mathrm{d}\tau} = -j_L(\tau) \iint_{-\infty}^{\infty} \frac{\mathrm{d}x\, \mathrm{d}y}{\pi w_0^2} n_{\mathrm{T}}(x, y, \tau) E_p(x, y, \tau) \mathrm{e}^{-\mathrm{i}\psi_p(x,y,\tau)} \langle \mathrm{e}^{-\mathrm{i}f\xi} \rangle_{\xi_0, \nu_0}, \tag{8.16}$$

which determines how each of the transverse modes p is driven by the electron beam.

8.3 The multimode pendulum equation

To determine the mode dependence of the pendulum equation from section 6.2, expand the optical field a in the analogous form of (8.7) as

$$a = |a| \mathrm{e}^{\mathrm{i}\phi} = \sum_p a_p E_p \mathrm{e}^{\mathrm{i}\psi_p} = \sum_p |a_p| E_p \mathrm{e}^{\mathrm{i}\psi_p} \mathrm{e}^{\mathrm{i}\phi_p}, \tag{8.17}$$

where a_p has the same conversion from c_p as a has from $\hat{E}$ (cf (8.14) and (6.19)). From the pendulum equation, (6.22),

$$\frac{\mathrm{d}\nu}{\mathrm{d}\tau} = |a| \cos(f\xi + \phi), \tag{8.18}$$

we have
$$\frac{\mathrm{d}\nu}{\mathrm{d}\tau} = |a| \frac{1}{2} \mathrm{e}^{\mathrm{i}f\xi} \mathrm{e}^{\mathrm{i}\phi} + |a| \frac{1}{2} \mathrm{e}^{-\mathrm{i}f\xi} \mathrm{e}^{-\mathrm{i}\phi} \tag{8.19}$$

$$= \frac{1}{2} a \mathrm{e}^{\mathrm{i}f\xi} + \frac{1}{2} a^* \mathrm{e}^{-\mathrm{i}f\xi} \tag{8.20}$$

$$= \frac{1}{2} \sum_p |a_p| E_p \mathrm{e}^{\mathrm{i}\psi_p} \mathrm{e}^{\mathrm{i}\phi_p} \mathrm{e}^{\mathrm{i}f\xi} + \frac{1}{2} \sum_p |a_p| E_p \mathrm{e}^{-\mathrm{i}\psi_p} \mathrm{e}^{-\mathrm{i}\phi_p} \mathrm{e}^{-\mathrm{i}f\xi}, \tag{8.21}$$

or
$$\left.\frac{\mathrm{d}\nu}{\mathrm{d}\tau}\right|_{x,y} = \sum_p |a_p| E_p(x, y, \tau) \cos(f\xi|_{x,y} + \phi_p + \psi_p(x, y, \tau)). \tag{8.22}$$

This is the *multimode pendulum equation*, which determines how the electrons are driven by the complete superposition of modes. Note that the phase space coordinates $\{\xi, \nu\}$ now depend on x, y through the (x, y)-dependence of the modes $E_p \exp \mathrm{i}\psi_p$. Physically this makes sense—we expect the phase space evolution to be different at the fringes of the beam where the field is weaker.

We should also probably comment here on the origin of the τ-dependence in the longitudinal current density $j_L(\tau)$. The total current at any co-moving point in the electron beam is a constant in the lab frame, being expressed as so many amps. However, the transverse dependence contained in n_T can change with τ along the undulator, say, if the electron beam is focused to a waist. The longitudinal factor j_L must therefore be compensated by a corresponding τ-dependence that keeps the total current constant.

The mode evolution equation and the multimode pendulum equation also conserve energy. Starting from the multimode pendulum equation in the form given by (8.21), write

$$\left\langle \frac{\mathrm{d}\nu}{\mathrm{d}\tau} \right\rangle_{\xi_0,\nu_0} = \frac{1}{2}\sum_p |a_p| \mathrm{e}^{\mathrm{i}\phi_p} E_p \mathrm{e}^{\mathrm{i}\psi_p} \langle \mathrm{e}^{\mathrm{i}f\xi} \rangle_{\xi_0,\nu_0} + \frac{1}{2}\sum_p |a_p| \mathrm{e}^{-\mathrm{i}\phi_p} E_p \mathrm{e}^{-\mathrm{i}\psi_p} \langle \mathrm{e}^{-\mathrm{i}f\xi} \rangle_{\xi_0,\nu_0}. \tag{8.23}$$

Multiply by $j_L n_\mathrm{T}/\pi w_0^2$ and integrate over x, y to obtain

$$\iint \frac{\mathrm{d}x\,\mathrm{d}y}{\pi w_0^2} j_L n_\mathrm{T} \left\langle \frac{\mathrm{d}\nu}{\mathrm{d}\tau} \right\rangle_{\xi_0,\nu_0} = \frac{1}{2}\sum_p a_p \iint \frac{\mathrm{d}x\,\mathrm{d}y}{\pi w_0^2} j_L n_\mathrm{T} E_p \mathrm{e}^{\mathrm{i}\psi_p} \langle \mathrm{e}^{\mathrm{i}f\xi} \rangle_{\xi_0,\nu_0}$$
$$+ \frac{1}{2}\sum_p a_p^* \iint \frac{\mathrm{d}x\,\mathrm{d}y}{\pi w_0^2} j_L n_\mathrm{T} E_p \mathrm{e}^{-\mathrm{i}\psi_p} \langle \mathrm{e}^{-\mathrm{i}f\xi} \rangle_{\xi_0,\nu_0} \tag{8.24}$$

or

$$-2\iint \frac{\mathrm{d}x\,\mathrm{d}y}{\pi w_0^2} j_L n_\mathrm{T} \left\langle \frac{\mathrm{d}\nu}{\mathrm{d}\tau} \right\rangle_{\xi_0,\nu_0} = \sum_p a_p \frac{\mathrm{d}a_p^*}{\mathrm{d}\tau} + \sum_p a_p^* \frac{\mathrm{d}a_p}{\mathrm{d}\tau} \tag{8.25}$$

from (8.16), or

$$\frac{\mathrm{d}}{\mathrm{d}\tau}\sum_p |a_p|^2 = -2\iint \frac{\mathrm{d}x\,\mathrm{d}y}{\pi w_0^2} j_L n_\mathrm{T} \left\langle \frac{\mathrm{d}\nu}{\mathrm{d}\tau} \right\rangle_{\xi_0,\nu_0}. \tag{8.26}$$

This is energy conservation, relating the energy gained in all the modes to the energy lost by the electrons. Indeed, we can insert the expressions for the dimensionless quantities to obtain, for a small volume ΔV,

$$\mathrm{d}\sum_p |c_p|^2 = -4\pi mc^2 \iint \frac{\mathrm{d}x\,\mathrm{d}y}{\pi w_0^2} n_e \langle \mathrm{d}\gamma \rangle_{\Delta V}; \qquad n_e = n_L n_\mathrm{T} \equiv \frac{N_e}{\Delta V}. \tag{8.27}$$

Now, note that

$$\iint \frac{\mathrm{d}x\,\mathrm{d}y}{\pi w_0^2} |\hat{E}|^2 = \iint \frac{\mathrm{d}x\,\mathrm{d}y}{\pi w_0^2} \left| \sum_p c_p E_p \mathrm{e}^{\mathrm{i}\psi_p} \right|^2 = \sum_p |c_p|^2 \quad \text{by orthonormality.} \tag{8.28}$$

Inserting this result into the lhs of (8.27) and equating integrands for arbitrary ΔV yields

$$d\left(\frac{1}{4\pi}|\hat{E}|^2\Delta V\right) = -N_e\left\langle d(\gamma mc^2)\right\rangle_{\Delta V}, \tag{8.29}$$

as in section 7.4.

8.4 The filling factor

In general, the implicit radial dependence of the factor $\langle e^{-if\xi}\rangle_{\xi_0,\nu_0}$ greatly complicates the evaluation of the transverse integral in the wave equation. Benson (1985) develops several approaches to deal with this problem. If the electron beam is sufficiently small compared to the optical beam, the simplest approach historically has been to neglect the radial dependence of the phase-space evolution in the equations of motion; at least, such an approximation makes their integration tractable.

To employ this approximation, the factor $\langle e^{-if\xi}\rangle_{\xi_0,\nu_0}$ is simply extracted from the transverse integral in the wave equation. The phase space coordinates ξ, ν are then represented by their average $\bar{\xi}$, $\bar{\nu}$ over the radial electron beam distribution in the pendulum equation. The analysis of this approach in this section is based on a numerical code developed by Benson (1985) and agrees with more rigorous approximations if the electron beam is small; we will show below that energy is conserved, so the approximation is also self-consistent.

First, the number of electrons in length dz is

$$N_e \equiv \frac{(I/e)_{\mathrm{MKS}}}{c}\,dz = \iint dx\,dy\,n_L(z)n_T(x,y)\cdot dz = n_L\,dz\iint dx\,dy\,n_T(x,y), \tag{8.30}$$

where I is the electron beam current in amperes and $e = 1.602 \times 10^{-19}$ coulombs. We define the electron beam area A_e to be

$$A_e \equiv \iint dx\,dy\,n_T(x,y). \tag{8.31}$$

Note that if n_T is a dimensionless form factor this integral has the dimensions of area. The factor n_L is then

$$n_L = \frac{(I/e)_{\mathrm{MKS}}}{cA_e}. \tag{8.32}$$

For example, if

$$n_T = \exp\left[-\frac{x^2}{w_x^2}\right]\exp\left[-\frac{y^2}{w_y^2}\right],$$

then $A_e = \pi w_x w_y$, and if w_x has a τ-dependence due to horizontal focusing of the electron beam, then the corresponding time dependence will appear in n_L.

The mode evolution equation (8.16) is simply

$$\frac{da_p}{d\tau} = -j_L\left[\iint_{-\infty}^{\infty}\frac{dx\,dy}{\pi w_0^2}n_T E_p e^{-i\psi_p}\right]\cdot\langle e^{-if\xi}\rangle_{\xi_0,\nu_0}. \tag{8.33}$$

We define the *filling factor* f_p for the pth optical mode as

$$f_p \equiv \iint_{-\infty}^{\infty}\frac{dx\,dy}{\pi w_0^2}n_T\,E_p e^{-i\psi_p}, \tag{8.34}$$

which is simply the overlap integral of the radial electron beam distribution with the pth optical mode. The mode evolution equation with filling factor is then

$$\frac{da_p}{d\tau} = -j_L f_p\langle e^{-if\bar{\xi}}\rangle_{\xi_0,\nu_0}. \tag{8.35}$$

To include the filling factor in the pendulum equation, we start with equation (8.21),

$$\left.\frac{d\nu}{d\tau}\right|_{x,y} = \frac{1}{2}\sum_p |a_p|E_p e^{i\psi_p}e^{i\phi_p}e^{if\xi} + \frac{1}{2}\sum_p |a_p|E_p e^{-i\psi_p}e^{-i\phi_p}e^{-if\xi}, \tag{8.36}$$

and take an average $\langle\ldots\rangle_{n_T}$ over the transverse electron beam distribution to obtain

$$\frac{d\bar{\nu}}{d\tau} \equiv \left\langle\frac{d\nu}{d\tau}\right\rangle_{n_T} = \sum_p |a_p|\frac{1}{2}\left[e^{if\bar{\xi}}e^{i\phi_p}\langle E_p e^{i\psi_p}\rangle_{n_T} + e^{-if\bar{\xi}}e^{-i\phi_p}\langle E_p e^{-i\psi_p}\rangle_{n_T}\right]. \tag{8.37}$$

Now,

$$\langle E_p e^{i\psi_p}\rangle_{n_T} \equiv \frac{\iint E_p e^{i\psi_p}\,n_T\,dx\,dy}{\iint n_T\,dx\,dy} = \frac{\pi w_0^2 f_p^*}{A_e} \equiv R_F f_p^*;\quad R_F \equiv \frac{\pi w_0^2}{A_e}, \tag{8.38}$$

where the parameter R_F is the ratio of the optical mode area at the waist to the area A_e of the electron beam, (8.31). Similarly, $\langle E_p e^{-i\psi_p}\rangle_{n_T} = R_F f_p$. The pendulum equation then becomes

$$\frac{d\bar{\nu}}{d\tau} = \sum_p |a_p|R_F|f_p|\cos(f\bar{\xi} + \phi_p - \measuredangle f_p), \tag{8.39}$$

where the phase space coordinates $\bar{\xi}$, $\bar{\nu}$ are understood to represent averages over the transverse electron beam distribution. This is the multimode pendulum equation with filling factor. Note that since $\nu = \frac{d\xi}{d\tau}$, an average for ν implies the same average for ξ. In subsequent analyses this averaging will be implicit and we will drop the overbar notation.

When numerically integrating the coupled equations of motion, (8.39) and (8.35) or (8.22) and (8.16), the equations must be integrated self-consistently. Thus, at each time step $\Delta\tau$, the mode evolution equation is used to calculate how the electron distribution contributes to the growth of each mode p and the pendulum equation is then used to determine how the entire superposition of modes drives the electron motion.

To show energy conservation, start with the pendulum equation with the average $\langle \ldots \rangle_{n_T}$ written explicitly:

$$\left\langle \frac{\mathrm{d}\nu}{\mathrm{d}\tau} \right\rangle_{n_T} = \sum_p |a_p| R_F \, |f_p| \, \frac{1}{2}\left[\mathrm{e}^{\mathrm{i}(f\xi + \phi_p - \angle f_p)} + \mathrm{e}^{-\mathrm{i}(f\xi + \phi_p - \angle f_p)} \right] \tag{8.40}$$

$$\left\langle \left\langle \frac{\mathrm{d}\nu}{\mathrm{d}\tau} \right\rangle_{n_T} \right\rangle_{\xi_0,\nu_0} = \frac{1}{2}\sum_p a_p R_F f_p^* \langle \mathrm{e}^{\mathrm{i}f\xi} \rangle_{\xi_0,\nu_0} + \frac{1}{2}\sum_p a_p^* R_F f_p \langle \mathrm{e}^{-\mathrm{i}f\xi} \rangle_{\xi_0,\nu_0} \tag{8.41}$$

$$\left\langle j_L \left\langle \frac{\mathrm{d}\nu}{\mathrm{d}\tau} \right\rangle_{n_T} \right\rangle_{\xi_0,\nu_0} = \frac{1}{2}\sum_p a_p R_F \left[j_L f_p^* \langle \mathrm{e}^{\mathrm{i}f\xi} \rangle_{\xi_0,\nu_0} \right] + \frac{1}{2}\sum_p a_p^* R_F \left[j_L f_p \langle \mathrm{e}^{-\mathrm{i}f\xi} \rangle_{\xi_0,\nu_0} \right] \tag{8.42}$$

$$= -\frac{1}{2} R_F \sum_p \left(a_p \frac{\mathrm{d}a_p^*}{\mathrm{d}\tau} + a_p^* \frac{\mathrm{d}a_p}{\mathrm{d}\tau} \right) \tag{8.43}$$

$$= -\frac{1}{2} R_F \sum_p \frac{\mathrm{d}|a_p|^2}{\mathrm{d}\tau} \tag{8.44}$$

$$= -\frac{1}{2} \frac{\pi w_0^2}{\iint n_T \, \mathrm{d}x \, \mathrm{d}y} \sum_p \frac{\mathrm{d}|a_p|^2}{\mathrm{d}\tau}. \tag{8.45}$$

Now, we formally interchange the order of averaging on the lhs; this is possible if $\{\xi_0, \nu_0\} \equiv \{\bar{\xi}_0, \bar{\nu}_0\}$ are independent of (x, y), as our approximation assumed from the start. We then get

$$-2\left\langle j_L \left\langle \frac{\mathrm{d}\nu}{\mathrm{d}\tau} \right\rangle_{\xi_0,\nu_0} \right\rangle_{n_T} \equiv -2 \frac{\iint \mathrm{d}x \, \mathrm{d}y \, n_T \, j_L \left\langle \frac{\mathrm{d}\nu}{\mathrm{d}\tau} \right\rangle_{\xi_0,\nu_0}}{\iint n_T \, \mathrm{d}x \, \mathrm{d}y}$$

$$= \frac{\pi w_0^2}{\iint n_T \, \mathrm{d}x \, \mathrm{d}y} \sum_p \frac{\mathrm{d}|a_p|^2}{\mathrm{d}\tau}, \tag{8.46}$$

$$\text{or} \quad -2 \iint \frac{\mathrm{d}x \, \mathrm{d}y}{\pi w_0^2} n_T \, j_L \left\langle \frac{\mathrm{d}\nu}{\mathrm{d}\tau} \right\rangle_{\xi_0,\nu_0} = \frac{\mathrm{d}}{\mathrm{d}\tau} \sum_p |a_p|^2, \tag{8.47}$$

which is the expression for energy conservation, (8.26), from section 8.3.

References

Benson S V 1985 Diffractive effects and noise in short-pulse free-electron lasers *PhD Thesis* Stanford University, Palo Alto, CA

Elleaume P and Deacon D A G 1984 Transverse mode dynamics in a free-electron laser *Appl. Phys.* B **33** 9–16

Siegman A E 1986 *Lasers* (Mill Valley, CA: University Science Books)

Chapter 9

Small-signal gain—first derivation

9.1 Gain from energy conservation

In the first application, we wish to calculate the fractional increase in optical power on a single pass through the undulator in the small signal regime. In the current derivation we will use energy conservation, demonstrated in section 7.4, by calculating the average energy lost by the electrons in the plane wave approximation and then equating this to the increase in optical energy. A second derivation in chapter 14 will proceed by integrating the coupled equations directly.

Start with the dimensionless FEL pendulum equation, (7.30), on the fundamental $f = 1$,

$$\frac{d\nu}{d\tau} = |a|\cos(\xi + \phi). \tag{9.1}$$

We obtain an exact integral for $\nu(\xi)$, i.e. ν as a function of ξ, for the case in which $|a|$ and ϕ are held constant, by multiplying both sides by $\nu = \frac{d\xi}{d\tau}$ and integrating from 0 to τ:

$$\nu\frac{d\nu}{d\tau} = |a|\cos(\xi + \phi) \cdot \frac{d\xi}{d\tau} \tag{9.2}$$

$$\frac{d}{d\tau}\left(\frac{1}{2}\nu^2\right) = \frac{d}{d\tau}(|a|\sin(\xi + \phi)) \tag{9.3}$$

$$\nu^2 - \nu_0^2 = 2|a|\big(\sin(\xi + \phi) - \sin(\xi_0 + \phi)\big). \tag{9.4}$$

We will use this equation to derive the gain to lowest order in $|a|$. However, our analysis has already employed another approximation—that both $|a|$ and ϕ remain constant. This, of course, is not the case; after all, our goal is to calculate the change in $|a|^2$ on a single pass. Therefore, our derivation will really only be valid for *small gain*, for which the fractional change in $|a|^2$ is sufficiently small.

doi:10.1088/978-1-6270-5573-4ch9

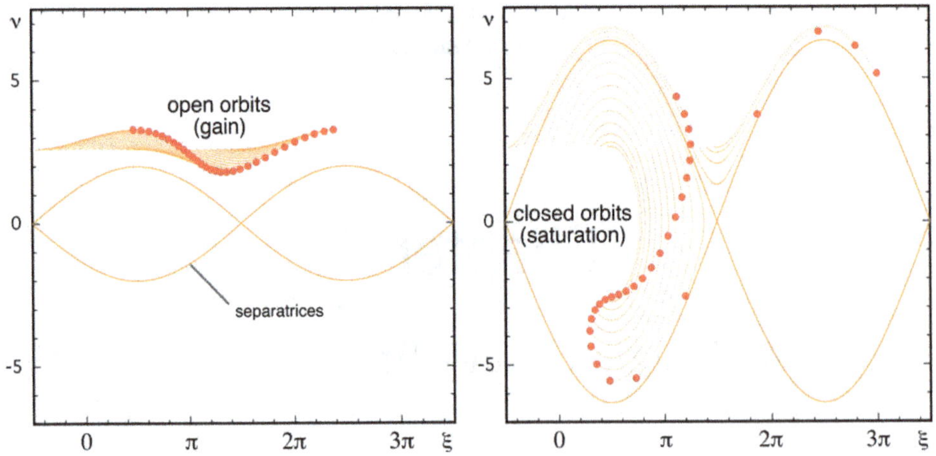

Figure 9.1. Phase space trajectories in constant optical fields. Left: $|a| = 1$; $\nu_0 = 2.606$. Right: $|a| = 10$; $\nu_0 = 2.606$. Bold dots show the final phase space distributions ($\tau = 1$).

The result is independent of the choice of ϕ; choose $\phi = 0$. Then the phase-space trajectories $\nu(\xi)$ are given by

$$\nu^2 - \nu_0^2 = 2|a|(\sin\xi - \sin\xi_0). \tag{9.5}$$

This equation is valid for any magnitude of $|a|$, even at saturation, as long as $|a|$ remains approximately constant. (This assumption is even more realistic in the saturated, large signal regime, where the fractional change in $|a|^2$ equals the typically small fractional cavity loss.) The phase-space trajectories for $\tau = 0 \to 1$ are as shown in figure 9.1.

The *separatrices* are the trajectories (ξ_s, ν_s) that separate the open and closed orbit regions; in the current example with $\phi = 0$, they contain the point $(\xi_0 = -\frac{\pi}{2}, \nu_0 = 0)$ and are given by

$$\nu_s = \pm\sqrt{2|a|}\sqrt{1 + \sin\xi_s}; \qquad \text{separatrix height} = \pm 2\sqrt{|a|}. \tag{9.6}$$

The actual trajectories followed by the electrons depend on the initial values (ξ_0, ν_0) and the magnitude of $\sqrt{|a|}$. If $\sqrt{|a|} \ll \nu_0$, as in the small signal regime at startup, then the electrons remain in the open orbit region and yield a large fractional gain. However, if $2\sqrt{|a|} > \nu_0$ then the electrons will be captured in the closed orbits and circulate to the bottom of the phase space 'buckets'. This behavior is the origin of saturation in the FEL: the electrons do not lose energy continuously, but instead fall only a finite distance $\Delta\nu$ before they come back up again, thus increasing their energy on average at the expense of the optical wave.

Anyway, let's continue with the small-signal gain calculation. In this case $\sqrt{|a|} \ll \nu_0$, so we use (9.5) to develop a perturbation expansion in the small parameter $|a|$, to order $|a|^2$, obtaining

$$\nu = \left[\nu_0^2 + 2|a|(\sin\xi - \sin\xi_0)\right]^{\frac{1}{2}} \tag{9.7}$$

$$= \nu_0\left[1 + \frac{|a|}{\nu_0^2}(\sin\xi - \sin\xi_0) - \frac{|a|^2}{2\nu_0^4}(\sin\xi - \sin\xi_0)^2 + - \cdots\right] \tag{9.8}$$

$$= \nu_0 + \frac{|a|}{\nu_0}(\sin\xi - \sin\xi_0)$$

$$- \frac{|a|^2}{2\nu_0^3}\left(\frac{1}{2} - \frac{1}{2}\cos 2\xi - 2\sin\xi\sin\xi_0 + \frac{1}{2} - \frac{1}{2}\cos 2\xi_0\right) + O\left(|a|^3\right), \tag{9.9}$$

where we expanded the square in the $|a|^2$ term in the third line and used some trig identities. We now integrate to obtain $\xi(\tau) = \int^{\tau} \nu \, d\tau'$ and substitute back into the above equation to obtain $\nu(\tau)$; but, to retain terms of order $|a|^2$ in $\nu(\tau)$, we will only need to find $\xi(\tau)$ to order $|a|$. So:

$$\xi(\tau) = \xi_0 + \nu_0\tau + \frac{|a|}{\nu_0}\int_0^{\tau}\left(\sin\xi(\tau') - \sin\xi_0\right)d\tau' + O\left(|a|^2\right) \tag{9.10}$$

$$= \xi_0 + \nu_0\tau + \frac{|a|}{\nu_0}\int_0^{\tau}\left(\sin(\xi_0 + \nu_0\tau' + O|a|) - \sin\xi_0\right)d\tau' + O\left(|a|^2\right) \tag{9.11}$$

$$= \xi_0 + \nu_0\tau + \frac{|a|}{\nu_0}\int_0^{\tau}\left(\sin(\xi_0 + \nu_0\tau') - \sin\xi_0\right)d\tau' + O\left(|a|^2\right) \tag{9.12}$$

$$= \xi_0 + \nu_0\tau - \frac{|a|}{\nu_0^2}\left[\cos(\xi_0 + \nu_0\tau) - \cos\xi_0 + \nu_0\tau\sin\xi_0\right] + O\left(|a|^2\right). \tag{9.13}$$

Now, substitute this equation for $\xi(\tau)$ into the expanded equation for $\nu(\tau)$, (9.9), to obtain (with more trig):

$$\nu - \nu_0 = \frac{|a|}{\nu_0}\left[\sin(\xi_0 + \nu_0\tau) - \cos(\xi_0 + \nu_0\tau)\frac{|a|}{\nu_0^2}\right.$$

$$\left.[\cos(\xi_0 + \nu_0\tau) - \cos\xi_0 + \nu_0\tau\sin\xi_0] - \sin\xi_0\right]$$

$$- \frac{|a|^2}{2\nu_0^3}\left[\frac{1}{2} - \frac{1}{2}\cos(2\xi_0 + 2\nu_0\tau) - 2\sin(\xi_0 + \nu_0\tau)\sin\xi_0 + \frac{1}{2} - \frac{1}{2}\cos 2\xi_0\right]$$

$$+ O\left(|a|^3\right). \tag{9.14}$$

Group terms in $|a|$, $|a|^2$ and use $\cos A \cos B = \cdots$; $\cos A \sin B = \cdots$; $\sin A \sin B = \cdots$:

$$\nu - \nu_0 = \frac{|a|}{\nu_0}[\sin(\xi_0 + \nu_0\tau) - \sin\xi_0] - \frac{|a|^2}{2\nu_0^3}\Big[1 + \cos(2\xi_0 + 2\nu_0\tau)$$
$$- \cos(2\xi_0 + \nu_0\tau) - \cos\nu_0\tau + \nu_0\tau(\sin(2\xi_0 + \nu_0\tau) - \sin\nu_0\tau)$$
$$+ 1 - \frac{1}{2}\cos(2\xi_0 + 2\nu_0\tau) + \cos(2\xi_0 + \nu_0\tau) - \cos\nu_0\tau - \frac{1}{2}\cos 2\xi_0\Big]$$
$$+ O\left(|a|^3\right). \tag{9.15}$$

Now, to calculate the average energy loss by all the electrons, we average over the initially uniform distribution in ξ_0. (We assume that ν_0 is the same for all electrons in the beam; the beam is then monoenergetic and $\langle\ldots\rangle_{\xi_0,\nu_0} \to \langle\ldots\rangle_{\xi_0}$.) The terms in $|a|$ drop out and the only surviving terms in $|a|^2$ are

$$\langle \nu - \nu_0 \rangle_{\xi_0} = -\frac{|a|^2}{2\nu_0^3}\left[2 - 2\cos\nu_0\tau - \nu_0\tau\sin\nu_0\tau\right]. \tag{9.16}$$

Note that the energy loss exhibits *lethargy*, or time dependence, which has important implications for dispersive and refractive effects in the electron beam. The energy loss equals zero at $\tau = 0$ before 'bunching' has occurred, but increases after a single pass through the undulator ($\tau = 1$) to a value of

$$\langle \nu - \nu_0 \rangle_{\xi_0} = -\frac{|a|^2}{2\nu_0^3}\left[2 - 2\cos\nu_0 - \nu_0\sin\nu_0\right]. \tag{9.17}$$

Recall from section 7.4 that

$$\Delta\nu = 4\pi N_w \frac{\Delta\gamma}{\gamma} \quad \text{and} \quad -n_e\, mc^2 \langle\Delta\gamma\rangle_{\xi_0} = \frac{\Delta|\hat{E}|^2}{4\pi}.$$ Therefore,

$$\langle \nu - \nu_0 \rangle_{\xi_0} = \langle \Delta\nu \rangle_{\xi_0} = 4\pi N_w \left\langle \frac{\Delta\gamma}{\gamma} \right\rangle_{\xi_0} = -\frac{N_w}{n_e\gamma mc^2}\Delta|\hat{E}|^2. \tag{9.18}$$

We substitute this expression for $\langle \nu - \nu_0 \rangle_{\xi_0}$ on the lhs of (9.17) and insert the full expression for the dimensionless field $|a|$ in terms of $|\hat{E}|$ on the rhs, to finally obtain the gain G:

$$-\frac{N_w}{n_e\gamma mc^2}\Delta|\hat{E}|^2 = -\frac{1}{2}\cdot\frac{16\pi^2 e^2 N_w^4 \lambda_w^2 \hat{K}_f^2}{\gamma^4 m^2 c^4}|\hat{E}|^2 \cdot \left[\frac{2 - 2\cos\nu_0 - \nu_0\sin\nu_0}{\nu_0^3}\right], \tag{9.19}$$

and thus
$$G \equiv \frac{\Delta|\hat{E}|^2}{|\hat{E}|^2} = \frac{8\pi^2 e^2 N_w^3 \lambda_w^2 \hat{K}_f^2}{\gamma^3 mc^2} n_e \cdot \left[\frac{2 - 2\cos\nu_0 - \nu_0\sin\nu_0}{\nu_0^3}\right] \tag{9.20}$$

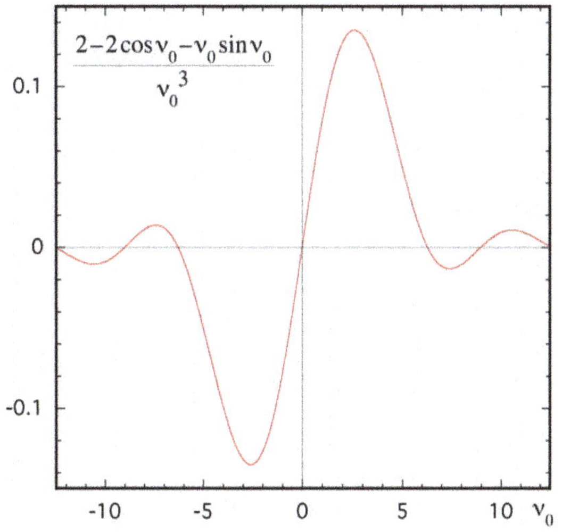

Figure 9.2. The FEL antisymmetric small-signal gain function. Maximum gain occurs at $\nu_0^{\text{opt}} = 2.606$.

$$G = j \cdot \left[\frac{2 - 2\cos\nu_0 - \nu_0 \sin\nu_0}{\nu_0^3} \right], \tag{9.21}$$

where j is the dimensionless current density from (7.28). Using the relation $\mathrm{d}\nu = -2\pi N_w \frac{\mathrm{d}\omega}{\omega r}$ (11.4), the reader can confirm that this curve corresponds to the ω-derivative of the spontaneous spectrum, (5.4), a result that derives from the quantum theory of the FEL (Madey 1971).

The gain function in brackets is antisymmetric in ν_0. There is no gain for an initial phase velocity of $\nu_0 = 0$ as we explained in the introduction (section 1.3). Instead, the maximum gain occurs for an initial phase velocity of $\nu_0^{\text{opt}} = 2.606$, at which point $G = 0.135\,j$. The gain function is plotted in figure 9.2.

All other quantities in the parameter j are then known, so we can calculate the maximum gain in a particular case. However, the electron density n_e requires some elaboration. We know the number of electrons N_e in a length δz of beam (8.30),

$$N_e = \frac{(I/e)_{\text{MKS}}}{c}\,\delta z, \tag{9.22}$$

where I is the peak current in amps, etc, but what is the volume to consider? Is n_e the peak n_e on axis, or averaged over the electron beam volume, or averaged over the optical mode volume, or what? Well, go back to energy conservation, written as

$$-n_e\left\langle \Delta(\gamma mc^2) \right\rangle_{\Delta V} = \frac{\Delta|\hat{E}|^2}{4\pi}. \tag{9.23}$$

If the electron beam is sufficiently narrow compared to the optical beam (the plane-wave approximation), then $|\hat{E}|$ appearing in this equation, which is the $|\hat{E}|$

that interacts with the electrons and determines their phase space evolution, is the electric field on axis. However, the electron energy loss contributes to the entire optical mode, not just the part near the axis. Indeed, if the mode is a fundamental Gaussian beam with a $1/e^2$-intensity radius of w_0, then the total mode energy equals the optical energy density on axis times the mode area $\frac{1}{2}\pi w_0^2$ times δz, where the optical intensity is $I = I_0 \exp(-2r^2/w_0^2)$. Therefore, we may write the total energy lost by the electrons, $-N_e\langle\Delta(\gamma mc^2)\rangle_{\Delta V}$, as

$$-\frac{(I/e)_{\mathrm{MKS}}}{c}\,\delta z\,\left\langle\Delta(\gamma mc^2)\right\rangle_{\Delta V} = \Delta\left(\frac{1}{4\pi}|\hat{E}|^2\cdot\frac{\pi w_0^2}{2}\,\delta z\right) \tag{9.24}$$

$$\text{or}\quad -\frac{(I/e)_{\mathrm{MKS}}}{c\,\dfrac{\pi w_0^2}{2}}\left\langle\Delta(\gamma mc^2)\right\rangle_{\Delta V} = \frac{\Delta|\hat{E}|^2}{4\pi}. \tag{9.25}$$

Comparing this result with (9.23) thus reveals that n_e is *averaged over the optical mode*:

$$n_e = \frac{(I/e)_{\mathrm{MKS}}}{cA_{\mathrm{opt}}},\qquad \text{where } A_{\mathrm{opt}} = \frac{\pi w_0^2}{2}. \tag{9.26}$$

The fundamental Gaussian mode actually converges and diverges through the undulator in accordance with the Gaussian beam formula $w^2 = w_0^2[1+\zeta^2]$, (8.3), where A_{opt} in (9.26) is the optical beam area at the waist. We therefore obtain a more accurate expression for the single-pass gain by calculating the average value of the electron density along the undulator. For a TEM_{00} mode of Rayleigh range z_{R} with its waist at the midpoint of the undulator, the result is

$$\langle n_e\rangle = \frac{(I/e)_{\mathrm{MKS}}}{c}\left\langle\frac{2}{\pi w^2}\right\rangle_{\mathrm{und}} = \frac{(I/e)_{\mathrm{MKS}}}{c}\frac{2}{\pi w_0^2}\cdot\frac{z_{\mathrm{R}}}{L_w}\int_{-\zeta_w}^{\zeta_w}\frac{d\zeta}{1+\zeta^2} = \frac{(I/e)_{\mathrm{MKS}}}{c}\frac{q_M}{A_{\mathrm{opt}}}, \tag{9.27}$$

$$\text{where } q_M \equiv \frac{1}{\zeta_w}\tan^{-1}\zeta_w \tag{9.28}$$

$$\text{and } \zeta_w \equiv \frac{L_w}{2z_{\mathrm{R}}}. \tag{9.29}$$

The parameter ζ_w is the normalized undulator half-length and the factor q_M is a *gain reduction factor* that accounts for the convergence and divergence of the optical mode. The maximum small-signal gain for a monoenergetic, filamentary electron beam is then

$$G = 0.135\,j;\qquad j = \frac{8\pi^2e^2N_w^3\lambda_w^2\hat{K}_f^2}{\gamma^3mc^2}\,\frac{(I/e)_{\mathrm{MKS}}}{c}\,\frac{q_M}{A_{\mathrm{opt}}}, \tag{9.30}$$

where all quantities are in CGS units except for $I_{\rm MKS}$ and $e_{\rm MKS}$. In practice, this expression is modified by the presence of finite electron beam emittance, high electron beam currents, finite electron energy spread and short-pulse slippage effects. Correction factors for these effects are calculated in chapter 10.

Example. Calculate the small-signal gain in an 'ideal' Mark III-class FEL, neglecting the gain perturbations just noted. Assume the following typical parameters for this laser:

(1) Peak current $I = 30$ A
(2) Electron beam kinetic energy $E_K = 42$ MeV
(3) Undulator parameter $\hat{K}^2 = 1.0$ ($\hat{K}_f^2 = 0.740$)
(4) Undulator period $\lambda_w = 2.3$ cm
(5) Number of periods $N_w = 47$
(6) Rayleigh range $z_{\rm R} \equiv \pi w_0^2/\lambda = 53$ cm

Solution. From item 2 we calculate $\gamma = 83.2$ and from 3 and 4 we calculate a fundamental wavelength of $\lambda = 3.323$ μm (4.16). From item 6, we calculate the optical mode radius and area to be $w_0 = 749$ μm and $A_{\rm opt} = 0.00881\,{\rm cm}^2$ and from 4 and 5 we have $q_M = 0.780$.

Inserting the above values into (9.30) yields a small-signal gain of $G = 1.17 = 117\%$. This result indicates that the optical power more than doubles on a single pass through the undulator. Of course, this magnitude of gain certainly violates our small gain condition $\Delta |a|^2 \ll |a|^2$. Nevertheless, the gain *is* high and the laser turns on robustly.

9.2 Gain-spread theorem

Consider the expansion in $|a|$ and $|a|^2$ for the energy loss $(\nu - \nu_0)$ from (9.15). We saw that the lowest order term in $|a|$ does not contribute to the gain, since the average $\langle ... \rangle_{\xi_0}$ drops out. However, the corresponding energy spread $(\nu - \nu_0)^2$ to lowest order $|a|^2$ is found to be

$$(\nu - \nu_0)^2 = \frac{|a|^2}{\nu_0^2}\left[\sin(\xi_0 + \nu_0\tau) - \sin\xi_0\right]^2 \tag{9.31}$$

$$= \frac{|a|^2}{\nu_0^2}\left[\frac{1}{2} - \frac{1}{2}\cos(2\xi_0 + 2\nu_0\tau) + \cos(2\xi_0 + \nu_0\tau) - \cos\nu_0\tau + \frac{1}{2} - \frac{1}{2}\cos 2\xi_0\right], \tag{9.32}$$

and this does *not* average to zero; for $\tau = 1$ we get

$$\left\langle(\nu - \nu_0)^2\right\rangle_{\xi_0} = |a|^2\left[\frac{1 - \cos\nu_0}{\nu_0^2}\right] \neq 0 \text{ in general.} \tag{9.33}$$

In fact, by differentiating with respect to ν_0, we find that

$$\frac{\mathrm{d}}{\mathrm{d}\nu_0}\left\langle(\nu - \nu_0)^2\right\rangle_{\xi_0} = |a|^2\left[\frac{\nu_0 \sin\nu_0 - 2(1 - \cos\nu_0)}{\nu_0^3}\right], \tag{9.34}$$

which is proportional to the small signal gain. By reference to (9.17) we may write

$$\frac{1}{2}\frac{\mathrm{d}}{\mathrm{d}\nu_0}\left\langle(\nu - \nu_0)^2\right\rangle_{\xi_0} = \langle \nu - \nu_0\rangle_{\xi_0}, \tag{9.35}$$

where it is understood that each side is expressed to order $|a|^2$. Since $\delta\nu \propto \delta\gamma$ for small energy changes $\delta\gamma$, we may write this result as

$$\frac{1}{2}\frac{\mathrm{d}}{\mathrm{d}\gamma_0}\left\langle(\gamma - \gamma_0)^2\right\rangle_{\xi_0} = \langle \gamma - \gamma_0\rangle_{\xi_0}. \tag{9.36}$$

This result illustrates the *gain-spread theorem*. It was first proved by Madey (1979) and was later shown by Kroll (1982) to hold generally. Its significance lies in the fact that the rhs is related to the laser gain, while the lhs is related to the energy spread induced by the amplification process and to the spontaneous power emitted by the electrons.

9.3 Approximate solution of the FEL equations

For practical numerical simulations of FELs, the modified Maxwell–Lorentz equations of motion obtained in section 8.4 (appropriately modified to include slippage) often serve as a rigorous and appropriate starting point. These equations were written

$$\frac{\mathrm{d}\nu}{\mathrm{d}\tau} = \sum_p |a_p| R_F |f_p| \cos(f\xi + \phi_p - \angle f_p) \tag{9.37}$$

$$\frac{\mathrm{d}a_p}{\mathrm{d}\tau} = -j_L f_p \langle \mathrm{e}^{-\mathrm{i}f\xi}\rangle_{\xi_0,\nu_0}, \tag{9.38}$$

and analytically incorporate the transverse dependence of the electron and optical beams to good approximation via the filling factor f_p, (8.34). In these equations, the parameters R_F, f_p and j_L appearing on the rhs generally depend on τ as the beams propagate through the undulator; of course, the expression for energy conservation obtained in (8.47) demonstrates the self-consistency of these equations at each time step, independent of the specific time dependence of the various parameters. However, even for the simplest time dependence, the above equations are analytically intractable and can only be solved numerically.

Nevertheless, exact analytic integration of the coupled equations of motion is possible, at least to lowest order in the optical field, if the spatial beam geometries are independent of τ, e.g. for collimated electron and optical beams. Such beams, of course, do not exist in free space. However, the Rayleigh parameters for realistic beams are in many cases sufficiently long that suitable approximations can be imposed on the equations of motion, leading to solutions that not only yield fundamental physical insight into the full range of FEL operation, but also provide useful and sensible estimates for many practical FEL parameters including small signal gain, gain reduction effects, and optical power at saturation. This section is devoted to the development of these approximations for the subsequent exploration of FEL operation in chapters 10 through 12.

Start with the modified Maxwell-Lorentz equations of motion, (9.37) and (9.38),

$$\frac{\mathrm{d}a_p}{\mathrm{d}\tau} = -j_L f_p \langle \mathrm{e}^{-\mathrm{i}f\xi} \rangle_{\xi_0,\nu_0} \tag{9.39}$$

$$\frac{\mathrm{d}\nu}{\mathrm{d}\tau} = |a_p| R_F |f_p| \cos(f\xi + \phi_p - \measuredangle f_p), \tag{9.40}$$

where we include only the lowest-order transverse mode $p = 0$ in the superposition; this makes the problem tractable and also reflects the spatial filtering properties of narrow undulator vacuum chambers. Make the following change of variables,

$$a' = a_p R_F |f_p|; \qquad \phi' = \phi_p, \tag{9.41}$$

$$\text{with} \quad \frac{\mathrm{d}a'}{\mathrm{d}\tau} = \frac{\mathrm{d}a_p}{\mathrm{d}\tau} R_F |f_p| + a_p \frac{\mathrm{d}}{\mathrm{d}\tau} R_F |f_p| \tag{9.42}$$

$$\text{or} \quad \frac{\mathrm{d}a'}{\mathrm{d}\tau} = \frac{\mathrm{d}a_p}{\mathrm{d}\tau} R_F |f_p| + a' \frac{\mathrm{d}}{\mathrm{d}\tau} \ln(R_F |f_p|). \tag{9.43}$$

Equations (9.39) and (9.40) are then written

$$\frac{\mathrm{d}a'}{\mathrm{d}\tau} = -j_L |f_p|^2 R_F \langle \mathrm{e}^{-\mathrm{i}(f\xi - \measuredangle f_p)} \rangle_{\xi_0,\nu_0} + a' \frac{\mathrm{d}}{\mathrm{d}\tau} \ln(R_F |f_p|) \tag{9.44}$$

$$\frac{\mathrm{d}\nu}{\mathrm{d}\tau} = |a'| \cos(f\xi - \measuredangle f_p + \phi'). \tag{9.45}$$

These equations contain no approximations besides the filling factor approximation of section 8.4. Analytic progress can be made on several fronts. First, for typical beam profiles, the second term on the rhs of (9.44) is odd in the normalized position ζ (8.4). Thus, in a symmetric resonator with its waist at the midpoint of the undulator, the second term drops out upon integration through the undulator if $a' \simeq$ const in a single pass. This condition is reasonably satisfied both in the small signal regime if the laser gain is small, where $\Delta |a'|^2 \ll |a'|^2$ by definition, and in the large signal regime at saturation, where $\Delta |a'|^2 / |a'|^2$ in steady state equals the typically small fractional cavity losses. For large fractional gains in the small signal regime, the second term does not generally integrate to zero. However, the solutions obtained by neglecting it agree surprisingly well with numerical simulations of the full coupled equations of motion, particularly if the beam divergence is not too large.

Based on the above considerations, the traditional ansatz employed at this point is to simply drop the second term in the wave equation, which is equivalent to assuming that geometric factor $R_F|f_p|$ is independent of τ. This condition is consistently imposed on the time-dependent geometric factor in the first term by ensuring that the resulting equations satisfy energy conservation. In absence of the second term in (9.44), instantaneous energy conservation takes the form (section 7.4, (7.34))

$$-2j_L|f_p|^2R_F\left\langle\frac{\mathrm{d}\nu}{\mathrm{d}\tau}\right\rangle_{\xi_0,\nu_0} = \frac{\mathrm{d}}{\mathrm{d}\tau}\left\{|a_p|^2R_F^2|f_p|^2\right\}, \tag{9.46}$$

where we substituted the expression for a' from (9.41). If $R_F|f_p|$ is to be independent of τ, as we wish to impose, then dividing both sides by $R_F^2|f_p|^2$ yields the geometric factor j_L/R_F on the lhs. From (8.15), (8.32) and (8.38), we see that this factor is also independent of τ, since the only remaining geometric factor A_e drops out. Consequently, all of the time dependence is contained in the dynamic variables ν and $|a_p|^2$.

Since the geometric factors on both sides of (9.46) can consistently be made time-independent, the second step in the ansatz is to replace them with their time-average during the interaction. The normalized current density j_F appearing on the lhs is defined as

$$j_F \equiv \langle j_L|f_p|^2R_F\rangle_{\mathrm{und}}, \tag{9.47}$$

and the optical field $|a'|^2$ appearing on the rhs scales as

$$|a'|^2 = |a_p|^2\langle R_F^2|f_p|^2\rangle_{\mathrm{und}}, \tag{9.48}$$

not, for example, as $a' = a_p\langle R_F|f_p|\rangle$, as (9.41) would imply. Energy conservation is written in the alternative forms

$$-2j_F\langle \mathrm{d}\nu\rangle_{\xi_0,\nu_0} = \mathrm{d}|a'|^2 \qquad \text{or} \qquad -2\frac{j_L}{R_F}\langle \mathrm{d}\nu\rangle_{\xi_0,\nu_0} = \mathrm{d}|a_p|^2. \tag{9.49}$$

For operation on the fundamental harmonic $f = 1$, which will be the subject of our investigations in chapters 10 and 11, we introduce the phase space variables

$$\xi' \equiv \xi - \measuredangle f_p \tag{9.50}$$

$$\nu' \equiv \frac{\mathrm{d}\xi'}{\mathrm{d}\tau} = \nu - \frac{\mathrm{d}}{\mathrm{d}\tau}\measuredangle f_p. \tag{9.51}$$

The equations of motion, (9.44) and (9.45), then assume the standard form

$$\frac{\mathrm{d}a'}{\mathrm{d}\tau} = -j_F\langle \mathrm{e}^{-\mathrm{i}\xi'}\rangle_{\xi_0',\nu_0'} \tag{9.52}$$

$$\frac{d\nu'}{d\tau} = |a'| \cos(\xi' + \phi'), \tag{9.53}$$

if we assume that the phase $\measuredangle f_p$ has only a first order τ-dependence. This is a sensible approximation in many cases, the specific consequences of which are explored in section 9.4. The small-signal gain in the small gain regime is then

$$G = 0.135\, j_F, \tag{9.54}$$

obtained by analogy with the calculation in section 9.1 by replacing $j \to j_F$. (Although we explicitly used only the pendulum equation in section 9.1 to calculate the gain, the result depended on the self-consistency of both of the FEL coupled equations, (7.30) and (7.31), as expressed by energy conservation. In chapter 14, we calculate the gain from both equations without recourse to energy conservation.)

Operation on higher harmonics f will be considered separately in chapter 12.

While the ansatz introduced in this section comprises a distinct mathematical problem from the problem posed by coupled equations (9.39) and (9.40)—which can only be solved numerically—it yields surprisingly accurate results within all regimes of FEL operation when compared to numerical simulations. But perhaps its greatest value is the physical insight it brings to the physics of FELs, including an elucidation of various gain perturbations in chapter 10 and a qualitative and quantitative understanding of laser saturation and output power in chapters 11 and 12. Finally, of course, the troublesome second term in (9.44) vanishes altogether in the limit of collimated electron and optical beams; all of our results in chapters 10 through 12 are rigorously valid in that limit.

Filamentary beam. To provide an initial benchmark for the approximation, consider a filamentary electron beam of zero emittance, for which the results from chapter 8 yield

$$n_T(x, y) = \delta^{(2)}(x, y) \tag{9.55}$$

$$A_e = \iint n_T \, dx \, dy = 1 \tag{9.56}$$

$$n_L \equiv \frac{(I/e)_{MKS}}{cA_e} = \frac{(I/e)_{MKS}}{c} \tag{9.57}$$

$$R_F \equiv \frac{\pi w_0^2}{A_e} = \pi w_0^2, \tag{9.58}$$

and

$$f_p = \iint \frac{dx\, dy}{\pi w_0^2} n_T \sqrt{\frac{2}{1+\zeta^2}}\, e^{-\frac{x^2+y^2}{w_0^2(1+\zeta^2)}}\, e^{-i\frac{(x^2+y^2)\zeta}{w_0^2(1+\zeta^2)}}\, e^{i\tan^{-1}\zeta}$$

$$= \frac{1}{\pi w_0^2}\sqrt{\frac{2}{1+\zeta^2}}\, e^{i\tan^{-1}\zeta}, \tag{9.59}$$

where $\zeta = z/z_R$, and we calculated the filling factor f_p from (8.34) using (8.3) and (8.4) with $p = 0$. Since n_L and j_L are constant, the current density j_F from (9.47) is

$$j_F = \langle j_L |f_p|^2 R_F \rangle_{\text{und}} \tag{9.60}$$

$$= j_L \frac{2}{\pi w_0^2} \left\langle \frac{1}{1 + \zeta^2} \right\rangle_{\text{und}} \tag{9.61}$$

$$= j_L \frac{q_M}{A_{\text{opt}}}, \tag{9.62}$$

where $A_{\text{opt}} = \pi w_0^2/2$ and $q_M \equiv \langle [1 + \zeta^2]^{-1} \rangle_{\text{und}}$ was originally defined in (9.28) *et seq.* The small-signal gain on the fundamental harmonic $f = 1$ is then

$$G = 0.135\, j_F; \qquad j_F = \frac{8\pi^2 e^2 N_w^3 \lambda_w^2 \hat{K}_f^2}{\gamma^3 mc^2} \frac{(I/e)_{\text{MKS}}}{c} \frac{q_M}{A_{\text{opt}}}, \tag{9.63}$$

where we substituted the expressions for j_L and n_L from (8.15) and (9.57). This is the same result, subject to the same analytic approximations, that we obtained in (9.30) for a filamentary electron beam using arguments of energy conservation.

9.4 Gouy phase shift

In the preceding calculation of small-signal gain based on (9.52) and (9.53), the maximum gain is formally obtained for an initial phase velocity of $\nu_0'^{\text{opt}} = 2.606$; see figure 9.2. From (9.51), the actual phase velocity is

$$\nu_0^{\text{opt}} = \nu_0'^{\text{opt}} + \frac{\text{d}}{\text{d}\tau} \measuredangle f_p, \tag{9.64}$$

where $\measuredangle f_p$ is given by the Gouy phase factor in (9.59). If the mode divergence is not too large, so that the undulator lies within the confocal region of the mode, we can approximate the τ-dependence of the phase factor by its linear term,

$$\measuredangle f_p = \tan^{-1}\zeta \simeq \zeta = 2\zeta_w \tau - \zeta_w; \qquad \tau = 0 \to 1, \tag{9.65}$$

where the last equality applies to a symmetric resonator and $\zeta_w \equiv L_w/2z_R$. Maximum gain thus occurs at a phase velocity of

$$\nu_0^{\text{opt}} = \nu_0'^{\text{opt}} + 2\zeta_w = 2.606 + \frac{L_w}{z_R}. \tag{9.66}$$

This increase in phase velocity ν is evidently the result of an apparent increase in both the optical wavelength and phase speed of light (measured along the axis of the undulator) within the confocal region of the optical mode, leading to a modification of the slippage condition in section 1.2. To see this, note that the full axial phase of the traveling TEM_{00} optical mode, including the Gouy phase shift on axis, is

$$\phi = kz - \omega t - \tan^{-1}\frac{z}{z_R} \simeq \left(k - \frac{1}{z_R}\right)z - \omega t \equiv k_{\rm eff}\, z - \omega t, \tag{9.67}$$

where $k_{\rm eff}$ is an effective propagation constant along the z-axis. The axial phase speed c_z and wavelength λ_z of the optical radiation measured along the z-axis are thus

$$c_z \equiv \frac{\omega}{k_{\rm eff}} = \frac{c}{1 - \dfrac{1}{kz_R}} \tag{9.68}$$

$$\lambda_z \equiv \frac{2\pi}{k_{\rm eff}} = \frac{\lambda}{1 - \dfrac{1}{kz_R}}. \tag{9.69}$$

The slippage condition governing stimulated emission of radiation of wavelength λ, given by (1.6) for axially propagating plane waves and electrons traveling at speed $\bar{v}_z$ along the axis, must be modified to read

$$(c_z - \bar{v}_z)\Delta t_w = \lambda_z \tag{9.70}$$

$$\left(\frac{c}{1 - \dfrac{1}{kz_R}} - \bar{v}_z\right)\frac{\lambda_w}{\bar{v}_z} = \frac{\lambda}{1 - \dfrac{1}{kz_R}} \tag{9.71}$$

$$\left(\frac{1}{\bar{\beta}_z} - 1 + \frac{1}{kz_R}\right)\lambda_w = \lambda \tag{9.72}$$

$$\left(1 - \bar{\beta}_z + \frac{1}{kz_R}\right)\lambda_w = \lambda, \tag{9.73}$$

where we multiplied by $\bar{\beta}_z$ in the last line and set $\bar{\beta}_z = 1$ in the terms containing the experimentally measured parameters z_R and λ_w. Since $\bar{\beta}_z$ is fixed, we see that the apparent increase in wavelength along the undulator axis due to the Gouy phase shift, which contributes a tiny fractional increase via the denominator of (9.69), is accompanied by much larger fractional increase (still $\ll 1$) in the actual radiation wavelength λ appearing in the numerator of (9.69), for which (9.73) yields

$$\Delta\lambda = \frac{\lambda_w}{kz_R} = \frac{\lambda}{k_w z_R}. \tag{9.74}$$

From the definition of the phase velocity, $\nu = L_w\left[(k + k_w)\bar{\beta}_z - k\right]$, we calculate

$$\Delta\nu = -L_w \Delta k(1 - \bar{\beta}_z) = +2\pi L_w \frac{\Delta\lambda}{\lambda^2}(1 - \bar{\beta}_z) \tag{9.75}$$

$$= +\frac{2\pi L_w}{\lambda^2}\frac{\lambda_w}{kz_R}\frac{\lambda}{\lambda_w} \tag{9.76}$$

$$= +\frac{L_w}{z_R}, \tag{9.77}$$

consistent with (9.66). For the Mark III parameters listed in chapter 9 ($N_w = 47$; $\lambda_w = 2.3$ cm; $z_R = 53$ cm), the shift is from $\nu_0 = 2.6 \rightarrow 4.6$, corresponding to a fractional shift of $\Delta\lambda/\lambda = 0.7\%$.

While the spectrum of stimulated emission into the TEM_{00} mode is shifted to longer wavelengths, the spontaneous spectrum experiences the same shift, because the wavelength of spontaneous radiation due to Thomson scattering increases with off-axis angle. This off-axis radiation fills an entire angular spectrum of plane waves within a Gaussian distribution of width $\theta \lesssim \theta_m = w_0/z_R$, where w_0 is the mode radius at the waist.

The spontaneous wavelength λ emitted at angle θ due to Thomson scattering is calculated by transforming the propagation four-vector $[\frac{\omega'}{c}; \mathbf{k}'] = [k'; k'\sin\theta', 0, k'\cos\theta']$ from the ERF into the laboratory frame, where the wavelength λ' in the ERF is given in (1.1) and γ_z is the gamma factor of the EF. The '0' and '3' (i.e. 't' and 'z') components of the transformed equations are

$$k = \gamma_z[k' + \beta_z k' \cos\theta'] \tag{9.78}$$

$$k\cos\theta = \gamma_z[k'\cos\theta' + \beta_z k']. \tag{9.79}$$

Dividing (9.79) by (9.78) and solving for $\cos\theta'$ yields

$$\cos\theta' = \frac{\cos\theta - \beta_z}{1 - \beta_z\cos\theta}, \tag{9.80}$$

and substitution of $\cos\theta'$ into (9.78) reduces to

$$k = \frac{k'}{\gamma_z(1 - \beta_z\cos\theta)}. \tag{9.81}$$

The corresponding relation between the wavelengths λ and $\lambda' = \lambda_w/\gamma_z$ is

$$\lambda = \lambda'\gamma_z(1 - \beta_z\cos\theta) = \lambda_w(1 - \beta_z\cos\theta) \tag{9.82}$$

$$\simeq \lambda_w(1 - \beta_z + \theta^2/2) \tag{9.83}$$

$$= \lambda_w \left(\frac{1 + \hat{K}^2}{2\gamma^2} + \frac{\theta^2}{2} \right), \tag{9.84}$$

$$\text{or} \quad \lambda = \frac{\lambda_w}{2\gamma^2}\left(1 + \hat{K}^2 + \gamma^2\theta^2\right), \tag{9.85}$$

where we substituted the expression for $1 - \beta_z$ from (4.15). The increase in wavelength of spontaneous radiation emitted into the Gaussian mode is obtained by substituting $\theta_m = w_0/z_R = \lambda/\pi w_0$ to account for the entire angular spectrum of plane waves, yielding

$$\Delta\lambda = \frac{\lambda_w}{2\gamma^2}\left(\gamma^2\theta_m^2\right) = \frac{\lambda_w}{2}\frac{\lambda^2}{\pi^2 w_0^2} = \frac{\lambda}{k_w z_R}, \tag{9.86}$$

which agrees with (9.74).

Evidently, the picture of stimulated emission that we described in section 1.2 applies to the Gaussian mode; the spontaneous and stimulated spectra both exhibit the same shift and satisfy the same slippage condition. Spontaneous radiation into the mode can thus act as a seed for the subsequent process of stimulated emission. This circumstance, of course, is required both by the quantum theory of the FEL, in which the gain spectrum for a given mode must correspond to the derivative of the spontaneous spectrum (Madey 1971), and by the theory of stimulated emission in quantum optics, in which the stimulated and spontaneous emission processes formally act identically on a given mode and are distinguished only by the initial number of quanta. In section 12.3, we show that the same FEL wave equation that describes stimulated emission into the TEM_{00} Gaussian mode also accounts quantitatively for spontaneous emission into that mode.

References

Kroll N M 1982 A note on the Madey gain-spread theorem *Physics of Quantum Electronics: Free-Electron Generators of Coherent Radiation* vol 8, S F Jacobs *et al* eds (Reading, MA: Addison-Wesley) pp 315–23

Madey J M J 1971 Stimulated emission of bremsstrahlung in a periodic magnetic field *J. Appl. Phys.* **42** 1906–13

Madey J M J 1979 Relationship between mean radiated energy, mean squared radiated energy and spontaneous power spectrum in a power series expansion of the equations of motion in a free-electron laser *Nuovo Cimento* **50** 64–88

Classical Theory of Free-Electron Lasers

A text for students and researchers

Eric B Szarmes

Chapter 10

Gain reduction and other effects

10.1 Electron beam emittance

In real systems, the electron beam is neither perfectly filamentary nor perfectly collimated. This leads to two gain reduction effects: a reduction due to the finite overlap with the optical beam, which can be quantified through incorporation of the filling factor, and a reduction due to an effective spread in longitudinal velocities, which mimics an energy spread. We consider the former effect in this section and the latter effect in section 10.3. We also restrict our discussion in both of chapters 10 and 11 to the fundamental harmonic, $f = 1$.

For a mode matched electron beam of finite emittance, following the analogous calculation for the filamentary beam in section 9.3, we take

$$n_{\mathrm{T}}(x, y) = \exp\left[-\frac{x^2}{w_x^2\left(1 + \dfrac{z^2}{\beta_x^2}\right)}\right]\exp\left[-\frac{y^2}{w_y^2}\right] \tag{10.1}$$

$$A_e = \iint n_{\mathrm{T}}\,\mathrm{d}x\,\mathrm{d}y = \pi w_x w_y\left[1 + \frac{z^2}{\beta_x^2}\right]^{\frac{1}{2}} \tag{10.2}$$

and

$$n_L \equiv \frac{(I/e)_{\mathrm{MKS}}}{cA_e}; \quad R_F \equiv \frac{\pi w_0^2}{A_e}. \tag{10.3}$$

The electron beam parameters $w_{x,y}$ and $\beta_{x,y}$ are related to the normalized emittance $\epsilon_{x,y}^n$ as defined in section 3.1, (3.8),

$$\pi w_x^2 = \epsilon_x^n \frac{\beta_x}{\gamma}; \quad \pi w_y^2 = \epsilon_y^n \frac{\beta_y}{\gamma}. \tag{10.4}$$

doi:10.1088/978-1-6270-5573-4ch10

Note that the vertical beam radius w_y in (10.1) is assumed to be collimated by the vertical focusing effect of the undulator field, in which case w_y is given by (3.43). We also assume that the electron beam is horizontally focused to match the optical mode, so that $\beta_x = z_R$. With these assumptions, the filling factor f_p (8.34) becomes a relatively simple product of Gaussian integrals that can be evaluated in closed form,

$$f_p = \iint \frac{dx\,dy}{\pi w_0^2}\, e^{-\frac{x^2}{w_x^2(1+\zeta^2)}}\, e^{-\frac{y^2}{w_y^2}} \sqrt{\frac{2}{1+\zeta^2}}\, e^{-\frac{x^2+y^2}{w_0^2(1+\zeta^2)}}\, e^{-i\frac{(x^2+y^2)\zeta}{w_0^2(1+\zeta^2)}}\, e^{i\tan^{-1}\zeta} \tag{10.5}$$

$$= \sqrt{2}\sqrt{1+\zeta^2}\left[\frac{w_0^2}{w_x^2} + 1 + i\zeta\right]^{-\frac{1}{2}}\left[\frac{w_0^2}{w_y^2}(1+\zeta^2) + 1 + i\zeta\right]^{-\frac{1}{2}} e^{i\tan^{-1}\zeta}. \tag{10.6}$$

Using this result together with (10.2) and (10.3), we calculate

$$n_L |f_p|^2 R_F = \frac{(I/e)_{\rm MKS}}{c}\frac{\pi w_0^2}{A_e^2}|f_p|^2 \tag{10.7}$$

$$= \frac{(I/e)_{\rm MKS}}{c}\frac{\pi w_0^2\; 2(1+\zeta^2)}{\pi^2 w_x^2 w_y^2(1+\zeta^2)}\left[\left(\frac{w_0^2}{w_x^2}+1\right)^2 + \zeta^2\right]^{-\frac{1}{2}} \left[\left(\frac{w_0^2}{w_y^2}\left[1+\zeta^2\right]+1\right)^2 + \zeta^2\right]^{-\frac{1}{2}} \tag{10.8}$$

$$= \frac{(I/e)_{\rm MKS}}{c}\frac{2}{\pi w_0^2}\left[\left(1+\frac{w_x^2}{w_0^2}\right)^2 + \frac{w_x^4}{w_0^4}\zeta^2\right]^{-\frac{1}{2}} \left[\left(1+\zeta^2+\frac{w_y^2}{w_0^2}\right)^2 + \frac{w_y^4}{w_0^4}\zeta^2\right]^{-\frac{1}{2}}. \tag{10.9}$$

If the divergence of the beams is sufficiently small ($|\zeta| \lesssim 1$) and if the electron beam is substantially narrower than the optical mode ($w_{x,y} \ll w_0$), then we may neglect the last term in each of the radicals in (10.9). Upon averaging over the undulator, we find the laser gain to be proportional to

$$\langle n_L |f_p|^2 R_F\rangle_{\rm und} = \frac{(I/e)_{\rm MKS}}{c}\frac{1}{A_{\rm opt}}\left[1+\frac{w_x^2}{w_0^2}\right]^{-1}\left[1+\frac{w_y^2}{w_0^2}\right]^{-\frac{1}{2}}\frac{1}{\zeta_w}\tan^{-1}\left\{\zeta_w\left[1+\frac{w_y^2}{w_0^2}\right]^{-\frac{1}{2}}\right\}, \tag{10.10}$$

where $\zeta_w = L_w/2z_R$ was originally defined in (9.29). We define the gain reduction factor q_E for emittance as

$$q_E = \left[1 + \frac{w_x^2}{w_0^2}\right]^{-1}\left[1 + \frac{w_y^2}{w_0^2}\right]^{-\frac{1}{2}} \frac{1}{\zeta_w}\tan^{-1}\left\{\zeta_w\left[1 + \frac{w_y^2}{w_0^2}\right]^{-\frac{1}{2}}\right\}. \qquad (10.11)$$

The small-signal gain in (9.63) is then modified in the case of finite emittance by replacing the gain reduction factor q_M by q_E, which accounts for both the divergence of the optical mode and the finite size of the electron beam; we see in particular that $q_E \to q_M$ as $w_{x,y} \to 0$, with $q_E < q_M$. The expression for the small-signal gain is

$$G = 0.135\, j_F; \quad j_F = \frac{8\pi^2 e^2 N_w^3 \lambda_w^2 \hat{K}_f^2}{\gamma^3 mc^2} \frac{(I/e)_{\text{MKS}}}{c} \frac{q_E}{A_{\text{opt}}}. \qquad (10.12)$$

For the conversion of optical power from $|a_p|^2$ to $|a'|^2$ which is needed in chapter 11, note from (10.3) that n_L and R_F are both proportional to A_e^{-1}. Thus, the factor $\langle R_F^2 |f_p|^2\rangle$ in (9.48) is evaluated by replacing $\frac{1}{c}(I/e)_{\text{MKS}} \to \pi w_0^2$ in the expression for $\langle n_L\, |f_p|^2 R_F\rangle$ in (10.10) above, and we immediately have

$$|a'|^2 = 2q_E\, |a_p|^2. \qquad (10.13)$$

10.2 High current and high gain

If $\Delta\, |a|^2 \simeq |a|^2$ on a single pass, then the analysis of chapter 9 is not valid and the explicit evolution of the optical field a must be included. We do this rigorously in chapter 14 by solving the coupled (7.30) and (7.31) for the small-signal gain in the high gain regime for a CW electron beam. The corresponding solution of (9.52) and (9.53) for a monoenergetic beam, with $j \to j_F$ formally, is

$$a'(\tau) = a_0' + \frac{\mathrm{i}j_F}{2}\int_0^\tau \mathrm{d}p \int_0^p \mathrm{d}q \cdot (p - q) \cdot a'(q) \cdot \mathrm{e}^{-\mathrm{i}\nu_0'(p-q)}. \qquad (10.14)$$

In the small gain regime for which $a'(q) \simeq a_0'$, we recover the previous expression for the small-signal gain, (9.21), with $j \to j_F$ in the current analysis, in addition to an imaginary term describing FEL dispersion. In the high gain regime with arbitrary current density j_F, numerical solution of (10.14) yields

$$G'(j_F) \equiv \frac{|a'(1)|^2 - |a_0'|^2}{|a_0'|^2}\bigg|_{\nu_0^{\text{opt}}} = 0.135\, j_F + 0.00486\, j_F^2 + 0.0000172\, j_F^3, \qquad (10.15)$$

with $\nu_0'^{\text{opt}} \simeq 2.606 - 0.022 j_F + 0.00016 j_F^2$. This expression for G' is accurate to $|\Delta G'|/G' < 0.04\%$ for $j_F \leqslant 22$. Nonlinear emission due to high current evidently increases the gain, even at lowest order in the optical field, and distorts and broadens the gain curve (figure 10.1). Note in particular that the small-signal gain is finite for $\nu_0' = 0$.

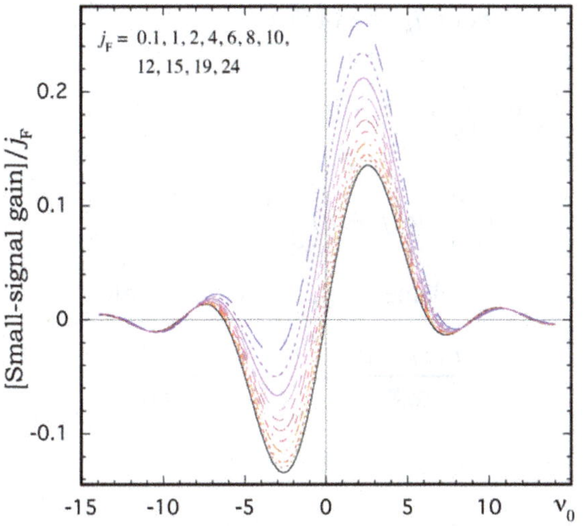

Figure 10.1. Distortion of the small-signal gain in the high current regime: $j_F = 0.1, \ldots, 24$.

In light of the symmetry of the phase space solutions (9.4) of the pendulum equation, (7.30), it may at first seem surprising that there should be any gain at resonance ($\nu_0 = 0$). However, while the phase space trajectories are intrinsically symmetric in ν (even as the amplitude $|a|$ increases), the wave equation (7.31), is not. The bunching imposed by the phase space revolutions is centered on the stable point at $\pi/2 = \xi + \phi_0$. The wave equation therefore contains a factor $\langle e^{-i\xi} \rangle \propto e^{-i(\pi/2-\phi_0)} = -ie^{i\phi_0}$, and the field increment $da \propto ie^{i\phi_0}$ is in quadrature with the initial field a; this is revealed in (10.14) by the factor of i in the integral term. Thus, while the separatrices grow symmetrically in amplitude as the field increases, they are simultaneously pulled to the *left*, breaking the symmetry of the phase space revolutions and yielding energy extraction from the electrons. This same discussion applies, of course, to (9.52) and (9.53).

The solution of (10.14) at $\nu_0' = 0$ can be readily developed in a power series expansion in j_F. The procedure is to formally substitute the solution for a' on the lhs into each successive appearance of a' in the integrand on the rhs. To second order in j_F this procedure yields

$$a'(\tau) = a_0 + \frac{ij_F}{2} \int_0^\tau dp \int_0^p dq \cdot (p - q) \cdot a'(q) \tag{10.16}$$

$$= a_0' + \frac{ij_F}{2} \int_0^\tau dp \int_0^p dq(p - q) \left[a_0' + \frac{ij_F}{2} \int_0^q dp' \int_0^{p'} dq'(p' - q')a_0' \right] + O\{j_F^3\}. \tag{10.17}$$

Collecting terms in powers of j_F, we obtain

$$\frac{a'(\tau) - a_0'}{a_0'} = +\frac{\mathrm{i}j_F}{2}\int_0^\tau \mathrm{d}p \int_0^p \mathrm{d}q(p-q)$$

$$-\frac{j_F^2}{4}\int_0^\tau \mathrm{d}p \int_0^p \mathrm{d}q(p-q)\int_0^q \mathrm{d}p' \int_0^{p'} \mathrm{d}q'(p'-q') + O\{j_F^3\}. \quad (10.18)$$

The successive integrals are elementary, and we calculate

$$\frac{a'(\tau) - a_0'}{a_0'} = \frac{\mathrm{i}j_F\tau^3}{12} - \frac{j_F^2\tau^6}{2880}. \quad (10.19)$$

To second order in j_F, the gain at $\nu_0 = 0$ is thus

$$G \equiv \frac{|a'(1)|^2 - |a_0'|^2}{|a_0'|^2} = \left|\frac{a'(1)}{a_0'}\right|^2 - 1 = \left|1 + \frac{\mathrm{i}j_F}{12} - \frac{j_F^2}{2880}\right|^2 - 1 \quad (10.20)$$

$$= \left(1 - \frac{j_F^2}{2880}\right)^2 + \left(\frac{j_F}{12}\right)^2 - 1 = \frac{j_F^2}{160}. \quad (10.21)$$

This solution is plotted in figure 10.2 for $j_F \leqslant 24$; it agrees with the numerical solution of (10.14) to within 0.6% and with a numerical integration of (9.52) and (9.53) (with j_F = const) to within 0.2%.

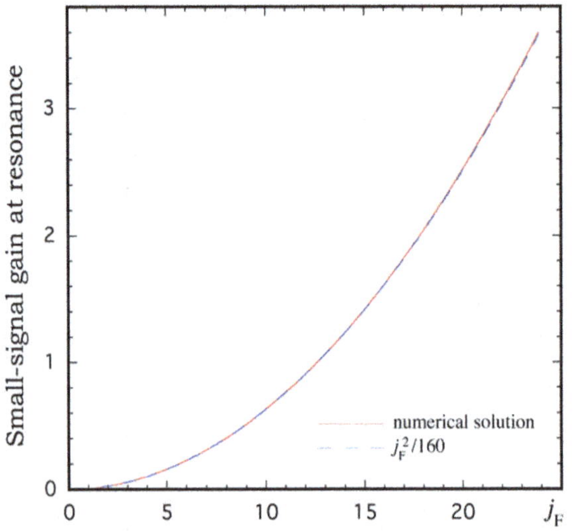

Figure 10.2. Small-signal gain at resonance ($\nu_0' = 0$) in the high current regime.

10.3 Energy spread

In an FEL, regardless of the presence or absence of energy spread, the frequency of the optical wave always evolves to the monochromatic value ω_0 yielding maximum gain. Larger values of energy spread then reduce the gain, because not all of the electrons enter the undulator with an optimum phase velocity of $\nu_0(\gamma^{\text{opt}}, \omega_0) = 2.606$. This corresponds to *inhomogeneous gain reduction.*

Small gain. In the small gain regime, energy spread can be incorporated as a straightforward extension of the energy-loss calculation in chapter 9. In particular, the result for $\langle \nu - \nu_0 \rangle_{\xi_0}$ in (9.17) applies to each subgroup of electrons within $[\nu_0, \nu_0 + d\nu_0]$ identified by their value of ν_0 in the overall distribution. By adding all of these contributions appropriately weighted by the number of electrons in each subgroup, we obtain the total energy lost by the electrons to the optical wave. Indeed, this procedure corresponds precisely to the prescription $\langle \ldots \rangle_{\nu_0}$. Let the energy spread be represented by a normalized Gaussian distribution centered on ν_p:

$$\rho(\nu_0) = \frac{1}{\sqrt{2\pi\sigma^2}} \exp\left[-\frac{(\nu_0 - \nu_p)^2}{2\sigma^2} \right]. \tag{10.22}$$

The gain reduction integral is then

$$\int \mathrm{d}\nu_0 \langle \nu - \nu_0 \rangle_{\xi_0} \rho(\nu_0)$$

$$\propto \int \mathrm{d}\nu_0 \frac{2 - 2\cos\nu_0 - \nu_0 \sin\nu_0}{\nu_0^3} \frac{1}{\sqrt{2\pi\sigma^2}} \exp\left[-\frac{(\nu_0 - \nu_p)^2}{2\sigma^2} \right] \equiv 0.135\, q_\gamma, \tag{10.23}$$

where the last equivalence defines the inhomogeneous gain reduction factor q_γ which multiplies the expression for the gain in (10.12). As laser oscillation evolves in the small signal regime, the optical frequency yielding maximum gain will settle on a phase velocity ν_0^{opt} centered on the peak of the distribution, ν_p. Numerical evaluation of (10.23) with $\nu_p = \nu_0^{\text{opt}} = 2.606$ then yields the approximation

$$q_\gamma \simeq \frac{1}{\left[1 + \left(\frac{\sigma}{3.27} \right)^2 \right]^{1.8}}, \tag{10.24}$$

which is accurate to $|\Delta q_\gamma|/q_\gamma < 1\%$ for $\sigma \leqslant 4$. The width σ of the Gaussian distribution is related to the full width $1/e$ energy spread $\delta\gamma_{1/e}$ by

$$\sigma = \sqrt{2}\, \pi N_w \left(\frac{\delta\gamma_{1/e}}{\gamma} \right). \tag{10.25}$$

High gain. To quantify the effect of inhomogeneous gain reduction for the case of high laser gain, it can be shown (see section 14.1) that the generalization of (10.14) to include a spread of energies, incorporated via $\langle \ldots \rangle_{\nu_0'}$, is

$$a'(\tau) = a_0' + \frac{\mathrm{i}j_F}{2}\int_0^\tau \mathrm{d}p \int_0^p \mathrm{d}q \cdot (p-q) \cdot a'(q) \cdot \langle \mathrm{e}^{-\mathrm{i}\nu_0'(p-q)} \rangle_{\nu_0'}. \tag{10.26}$$

For a Gaussian distribution of energies, (10.22), the result is

$$a'(\tau) = a_0' + \frac{\mathrm{i}j_F}{2}\int_0^\tau \mathrm{d}p \int_0^p \mathrm{d}q \cdot (p-q) \cdot a'(q) \cdot \mathrm{e}^{-\mathrm{i}\nu_p(p-q)}\mathrm{e}^{-\sigma^2(p-q)^2/2}. \tag{10.27}$$

For small j_F we recover the inhomogeneous gain reduction factor q_γ obtained from the energy-loss calculation, (10.24). For large j_F, with $\nu_p = \nu_0'^{\mathrm{opt}}$ given in the expression following (10.15), we find that the small-signal gain is closely approximated by

$$G_\gamma = G'(j_F q_\gamma), \tag{10.28}$$

where q_γ is given in (10.24) and $G'(\ldots)$ as a function of its argument is defined by (10.15). This expression is accurate to $|\Delta G_\gamma|/G_\gamma < 4\%$ for $\sigma \leqslant 4$ and $j_F \leqslant 22$; a more accurate prescription is given in section 11.6. (In contrast, the prescription $G_\gamma \to q_\gamma G'(j_F)$ yields an error $|\Delta G_\gamma|/G_\gamma > 50\%$ over the same range.)

Inhomogeneous gain reduction also includes the second effect of transverse emittance, even in an otherwise monoenergetic electron beam, because diverging or converging trajectories in a focused electron beam exhibit a spread of longitudinal velocities v_z in a single-sided distribution, corresponding to an effective energy spread. However, typical electron beams have sufficiently small emittance that this gain reduction effect is generally negligible. For an electron traveling at a small angle θ with azimuth φ about the axis of the undulator, we have, following the analysis in section 4.2,

$$\begin{aligned}\frac{1}{\gamma^2} &= 1 - (\beta_x^2 + \beta_y^2 + \beta_z^2) \\ &= 1 - \left[\left(-\frac{K}{\gamma}\cos k_w z + \bar{\beta}_z\theta\cos\varphi\right)^2 + \left(\bar{\beta}_z\theta\sin\varphi\right)^2 + \beta_z^2\right]\end{aligned} \tag{10.29}$$

so that

$$\begin{aligned}\beta_z^2 = 1 &- \frac{1}{\gamma^2} - \frac{K^2}{\gamma^2}\cos^2 k_w z + 2\frac{K}{\gamma}\bar{\beta}_z\theta\cos\varphi\cos k_w z \\ &- \bar{\beta}_z^2\theta^2\cos^2\varphi - \bar{\beta}_z^2\theta^2\sin^2\varphi\end{aligned} \tag{10.30}$$

$$\beta_z \simeq 1 - \frac{1}{2\gamma^2} - \frac{K^2}{2\gamma^2}\cos^2 k_w z + \frac{K}{\gamma}\bar{\beta}_z\theta\cos\varphi\cos k_w z - \frac{1}{2}\bar{\beta}_z^2\theta^2 \tag{10.31}$$

$$\bar{\beta}_z = 1 - \frac{1+\hat{K}^2}{2\gamma^2} - \frac{\theta^2}{2}. \tag{10.32}$$

The final expression for $\bar{\beta}_z$ was obtained by averaging over magnet periods and setting $\bar{\beta}_z \simeq 1$ in the last term. The corresponding effective energy spread due to propagation at angle θ is then $\frac{d\gamma}{\gamma} = \frac{d\gamma_z}{\gamma_z} \equiv \gamma_z^2\, d\bar{\beta}_z =, \gamma^2\theta^2/2(1 + \hat{K}^2)$. As an example, in the Mark III FEL with $\gamma = 90$, $\hat{K} = 1$, $\epsilon_x^n = 8\pi$ mm · mrad, and (notation!) $\beta_x = z_R = 53$ cm, we have $w_x = 217$ µm and $\theta_m = 0.41$ mrad. The effective energy spread is $\frac{d\gamma}{\gamma} = 0.034\%$, which corresponds to $\sigma = 0.071$ and $q_\gamma = 0.9992$.

10.4 Short-pulse effects

When the laser is driven not by a CW electron beam but instead by short electron bunches, say, from an RF-linac, the laser gain is reduced due to slippage of the optical pulse past the electrons, especially for electron bunches for which the slippage length $N_w\lambda$ is a substantial fraction of the bunch length. In such cases, the longitudinal overlap cannot be maintained between the electron bunch and optical pulse within a single pass through the undulator.

When combined with laser lethargy (see (9.16)), this longitudinal walkoff also produces an interesting and important physical effect: since the time-dependent gain is higher at the downstream end of the undulator, the trailing end of the optical pulse experiences greater net amplification than the leading end (which has already slipped ahead of the electrons). As a consequence, the FEL interaction reduces the group velocity of the optical pulse, because the centroid of the pulse is shifted towards its trailing end in a single pass. To compensate this effect when operating the laser, the optical cavity length must be shortened by some 'cavity detuning' δL_c from its free-space synchronous length $L_0 = NcT_e/2$, where T_e is the repetition period of the electron bunches and N is an integer equal to the number of circulating optical pulses in the resonator.

The reduction in group velocity is a manifestation of the refractive index of the electron beam. We discuss dispersion of the electron beam from a different perspective in sections 14.2 and 15.4.

Dattoli *et al* 1993 carried out the complete theoretical analysis of this phenomenon and obtained exact quantitative solutions for the case of Gaussian electron bunch shapes. Unfortunately, several features of this *supermode theory* complicate its application in this discussion. The first is that the analysis is explicitly restricted to the small gain regime and thus cannot provide a rigorous description of pulse propagation for high laser gains. The second is that for typical RF-linac FELs, the actual electron bunches are generally not Gaussian in shape. Third, as the laser oscillation evolves in the small-signal regime, the optical pulses do not necessarily have time to converge to a single supermode (i.e. a stable pulse) before the laser saturates. The laser oscillation includes the effects of multiple supermodes, especially near the synchronous length where the supermode degeneracy is greater and the actual gain is higher than predicted by the supermode theory. Despite these issues, the basic results have proved useful in comparison with actual simulations, so we apply them tentatively in this discussion.

Here is a summary of results (Dattoli *et al* 1993). Define the *detuning parameter* ρ and *slippage parameter* μ_c,

$$\rho = \frac{4\,\delta L_c}{j_R\, N_w \lambda}; \qquad \mu_c = \frac{N_w \lambda}{\sigma_z}, \tag{10.33}$$

where σ_z characterizes the electron bunch length via $I = I_0 \exp\left(-\frac{z^2}{2\sigma_z^2}\right)$. The factor j_R appearing in ρ is given by $j_R = G_\gamma/0.135$. Then the gain of the stable circulating optical pulse (the 'fundamental supermode') is multiplied by the gain reduction factor

$$\mathcal{G}(\rho, \mu_c) = \frac{\rho}{\rho_s}\left\{1 - \ln\left[\frac{\rho}{\rho_s}\left(1 + \frac{\mu_c}{3}\right)\right]\right\}, \tag{10.34}$$

where $\rho_s = 0.456[\frac{0.135}{0.85}] = 0.0724$. The gain is maximized at a cavity detuning δL_c^p corresponding to

$$\rho_p = \frac{\rho_s}{1 + \frac{\mu_c}{3}}, \quad \text{i.e.} \quad \delta L_c^p = \frac{0.0181\, j_R\, N_w \lambda}{1 + \frac{\mu_c}{3}}, \tag{10.35}$$

and the maximum gain is

$$G_{ss}^p = G_\gamma\, \mathcal{G}\left(\rho_p, \mu_c\right) = 0.135\, j_R \cdot \frac{1}{1 + \frac{\mu_c}{3}}. \tag{10.36}$$

The gain of the stable optical pulse falls to zero for detunings of $\rho = 0$ and $\rho = e\rho_s/(1 + \frac{\mu_c}{3})$, the latter being a factor of $e = 2.71828$ greater than the optimum detuning in (10.35). This yields the width of the cavity detuning curve, i.e. the range δL_{cavdet} within which the laser gain is positive and the laser is generally able to turn on. Note that we have

$$\delta L_{\text{cavdet}} = \frac{e(0.0181\, j_R\, N_w \lambda)}{1 + \frac{\mu_c}{3}} = 0.364\, \frac{0.135\, j_R}{1 + \frac{\mu_c}{3}}\, N_w \lambda = (0.364\, N_w \lambda)\, G_{ss}^p. \tag{10.37}$$

Therefore, a measurement of δL_{cavdet} often facilitates a quick and sensible estimate of the laser gain G_{ss}^p.

10.5 Summary

In summary, we adopt the following prescription for estimating the gain in real systems:

1) Calculate

$$j_F = \frac{8\pi^2 e^2 N_w^3 \lambda_w^2 \hat{K}_f^2}{\gamma^3 m c^2}\, \frac{(I/e)_{\text{MKS}}}{c}\, \frac{q_E}{A_{\text{opt}}}; \qquad \text{(see (10.11) for } q_E\text{)}.$$

2) Calculate the CW laser gain, $G_\gamma = G'(j_F q_\gamma)$; (see (10.24) for q_γ and (10.15) for G').
3) Calculate the gain reduction due to short-pulse effects at any desired position on the cavity detuning curve, $G_{ss} = G_\gamma \mathcal{G}(\rho, \mu_c)$; (see (10.34) for $\mathcal{G}(\rho, \mu_c)$).

Example. How is the small-signal gain modified in the example at the end of section 9.1 if, in addition to the parameters listed there, we have a fractional $1/e$ energy spread of 0.3%, normalized emittances of $\epsilon_x^n = 8\pi$ mm · mrad and $\epsilon_y^n = 4\pi$ mm · mrad, mode matched β-functions of $\beta_x = z_R$ and $\beta_y = \gamma/\hat{K}k_w$ (3.39), and a full-width bunch length of $2\sigma_z/c = 2$ ps? Assume operation at the peak of the cavity detuning curve.

Solution. We calculate the following parameters:

i) $\sigma = 0.626$ and $q_\gamma = 0.937$ (10.25) and (10.24);
ii) $\beta_x = 53$ cm and $\beta_y = 30.5$ cm (3.39);
iii) $w_x = 226$ μm and $w_y = 121$ μm (10.4);
iv) $q_E = 0.700$ ((10.11), with $\zeta_w = 1.020$ and $w_0 = 749$ μm);
v) $\mu_c = N_w\lambda/\sigma_z = N_w\lambda/c/(1\text{ ps}) = 0.521$ ((10.33), with $\lambda = 3.323$ μm);
vi) $\mathcal{G}(\rho_p, \mu_c) = 0.852$ (peak of the cavity detuning curve, (10.36)).

Proceeding through the gain calculation then yields

1) $j_F = 7.792$;
2) $G_\gamma = 1.251$;
3) $G_{ss}^p = 1.07 = 107\%$.

The corresponding width of the cavity detuning curve indicated by (10.37), assuming the cavity losses are small, is $\delta L_{\text{cavdet}} = 61$ μm. This gain and detuning width are quite typical for the Mark III FEL. Usually, the cavity detuning is set to perhaps $\rho = \rho_p/3$, because this yields better energy extraction and optical power at saturation. For this detuning (10.34) yields

$$\mathcal{G}\left(\frac{\rho_p}{3}, \mu_c\right) = \frac{\frac{1}{3}}{1 + \frac{\mu_c}{3}}\left\{1 - \ln\left(\frac{1}{3}\right)\right\} = 0.596, \tag{10.38}$$

for which $G_{ss} = 0.746 = 75\%$. This is again a typically measured gain for configurations in which the electron bunch length is on the order of 2 ps.

It is interesting to reflect on the physical origin of the improvement in the saturated optical power near the synchronous length. As noted in section 10.4, the increase in gain for shorter cavity lengths is related to group delay and dispersion in the FEL interaction, which is ultimately related to lethargy, or time dependence, in the development of gain along the undulator. The small-signal gain is improved in the shortened cavity because the more slowly propagating optical pulses are better overlapped with the injected electron bunches from pass to pass.

However, laser lethargy and electron beam dispersion are absent in the saturated regime. The large optical fields immediately capture the electrons in closed phase space orbits, essentially eliminating the time dependence of the gain. Furthermore, the reduced fractional gain at saturation yields a smaller perturbation on the propagating optical pulse. Therefore, the saturated optical pulses exhibit a group velocity much closer to the vacuum speed of light. Consequently, at the shortened cavity lengths for which the small-signal gain is maximized, the optical pulses at saturation are now pushed increasingly *ahead* of the injected electron bunches on every pass, thereby reducing their physical overlap and overall energy extraction.

The optimal compromise between small-signal gain (with good pulse overlap in the small signal regime) and high power extraction (with good pulse overlap at saturation) is realized with a substantially smaller cavity detuning than the detuning that maximizes the small-signal gain.

Reference

Dattoli G, Giannessi L, Renieri A and Torre A 1993 Theory of Compton free electron lasers *Progress in Optics* vol 31, ed E Wolf (Amsterdam: North-Holland)

Chapter 11

Laser saturation and output power

11.1 The nature of FEL saturation

The essential richness of FEL physics—the complex, even chaotic, phase space evolution; frequency pulling effects; synchrotron oscillations and sideband formation; the high FEL efficiency; and not least, the MW-level optical micropulse powers delivered at GHz repetition rates—occurs at saturation.

To get at the nature of FEL saturation, examine the (ξ', ν') phase space trajectories for various magnitudes of optical field a', as shown in figure 11.1. These figures were calculated for $\tau = 0 \to 1$ by straightforward numerical integration of the coupled equations of motion, (9.52) and (9.53); the simulations used $j_F = 1$, but the trajectories are qualitatively independent of j_F.

In all the panels of figure 11.1, the electrons enter the undulator with fixed energy. The optical wave, which typically turns on from weak spontaneous radiation, evolves coherently with an optical frequency corresponding to the peak of the gain curve, with an initial phase velocity of $\nu_0' = 2.606$ for the electrons. This initial phase velocity remains basically fixed at the start of each pass as the coherent optical wave circulates within the resonator. The magnitude of the optical field determines the height of the separatrices, equal to $\Delta\nu' = \pm 2\sqrt{|a|}$ (see (9.6)). Saturation occurs when the separatrices, which initially have a height much smaller than ν_0', grow so large that they capture the electrons in closed phase space orbits.

Initially, in the small signal regime with $|a'|^2 \ll 1$, the electrons evolve in open orbits; the average energy loss of the electrons is small, but the fractional growth of the optical energy on a single pass, $\Delta|a'|^2/|a'|^2$, is large. As the optical field grows to the intermediate regime, $2\sqrt{|a'|} \sim \nu_0'$, some of the electrons become trapped in the phase space buckets enclosed by the separatrices; this represents the transition between the small signal and large signal regimes. In the large signal regime with $|a'|^2 \gg 1$, essentially all of the electrons are trapped in closed orbits within the phase space buckets. In these closed orbits, the electrons do not lose energy indefinitely; instead, they fall only as far as the bottom of the buckets before coming up again and

doi:10.1088/978-1-6270-5573-4ch11

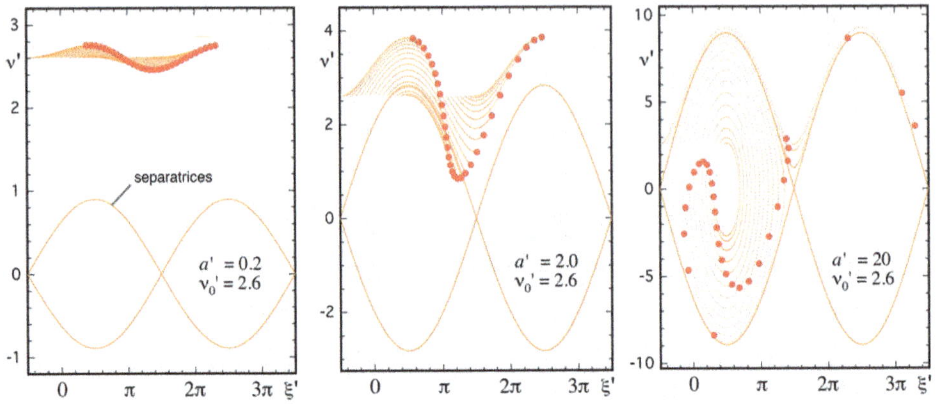

Figure 11.1. The nature of FEL saturation. Left: open orbits, high gain; $|a'|^2$ small, $\frac{\Delta|a'|^2}{|a'|^2}$ large. Center: intermediate regime, onset of particle trapping. Right: closed orbits, saturation; $|a'|^2$ large, $\frac{\Delta|a'|^2}{|a'|^2}$ small.

taking energy from the optical wave. This is the FEL saturation mechanism. In this regime, the energy loss of the electrons is large, but the fractional growth of the optical power, $\Delta|a'|^2/|a'|^2$, is small and continues to decrease with increasing $|a'|^2$. The optical field continues to grow until $\Delta|a'|^2/|a'|^2$ on a single pass decreases to a value equal to the fractional cavity loss. The laser then reaches a type of quasi-stable equilibrium, except for dynamic effects and the possible onset of instabilities discussed below.

11.2 Strong-saturation effects

Frequency pulling. In contrast to lasers based on atomic media, the peak of the FEL gain curve is not centered on the spontaneous spectrum. This has important implications for the evolution of the optical frequency in an FEL. If we compute the gain curves at various levels of saturation by self-consistent integration of the coupled Maxwell–Lorentz equations of motion, we find that in addition to the reduction in gain noted in the preceding section, there is a distortion of the gain curve that pulls peak the of the curve to higher values of $\nu_0'^{\text{opt}}$. This effect is illustrated in figure 11.2, generated for a small-gain current of $j_F = 0.1$.

Since the injected electron energy is typically held fixed, this increase in the peak value of ν_0' causes the optical frequency to be pulled to lower frequencies and longer wavelengths. Indeed, recall that the phase velocity defined in (6.8), $\nu = L_w[(k+k_w)\bar{\beta}_z - k]$, is an implicit function of both γ and ω, i.e. $\nu = \nu(\bar{\beta}_z(\gamma), k(\omega))$. We therefore have

$$d\nu = \frac{\partial \nu}{\partial \gamma} d\gamma + \frac{\partial \nu}{\partial \omega} d\omega \tag{11.1}$$

$$= \frac{\partial \nu}{\partial \bar{\beta}_z} \frac{d\bar{\beta}_z}{d\gamma} d\gamma + \frac{\partial \nu}{\partial k} \frac{dk}{d\omega} d\omega \tag{11.2}$$

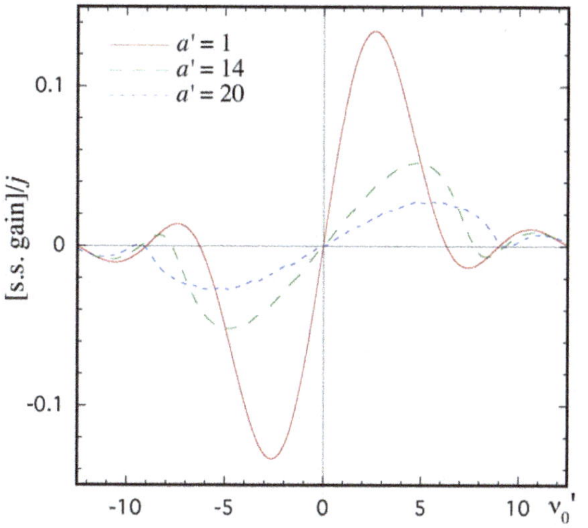

Figure 11.2. Distortion of the FEL gain function for increasing fields ($j_F = 0.1$).

$$d\nu = L_w k\,\frac{2k_w}{\gamma k}d\gamma + L_w\left(\bar{\beta}_z - 1\right)\frac{1}{c}d\omega \tag{11.3}$$

$$= 4\pi N_w\frac{d\gamma}{\gamma} - 2\pi N_w\frac{d\omega}{\omega}, \tag{11.4}$$

where we dropped $k_w \ll k$ in the expression for $\partial\nu/\partial\bar{\beta}_z$ and used the expression for $d\bar{\beta}_z/d\gamma$ from (6.14) in the third line, and inserted the slippage condition $1 - \bar{\beta}_z = \lambda/\lambda_w$ from (1.5) in the fourth line. In the evolution of an optical pulse of finite spectral width, whose initial wavelength is established via coherent evolution at the peak of the gain curve in the small-signal regime, the spectral content is slowly replaced in the large-signal regime by longer wavelengths due to the shift in $\nu_0'^{\text{opt}}$. This process occurs over many passes in the resonator as the old radiation decays away and is replaced by optical growth at the new wavelengths.

Frequency pulling is associated with the phase space behavior of the electrons at saturation, as we see in figure 11.3. In the presence of large optical fields, the phase space bucket height grows much higher than the initial phase velocity of $\nu_0'^{\text{opt}} = 2.606$. For separatrices corresponding to lower optical frequencies (larger values of ν_0'), the electrons have more room at the top of the buckets to fall a greater distance in $\Delta\nu'$ even as they remain trapped, yielding a greater energy loss and increased optical gain. The optical wave is thus pulled to lower frequencies and longer wavelengths over many passes in the resonator.

Synchrotron oscillations and sideband instability. In strong optical fields, the electrons can execute many closed orbit revolutions in a single pass through the undulator. These phase space revolutions are called *synchrotron oscillations* and the corresponding frequency is the *synchrotron frequency*. The revolutions are well

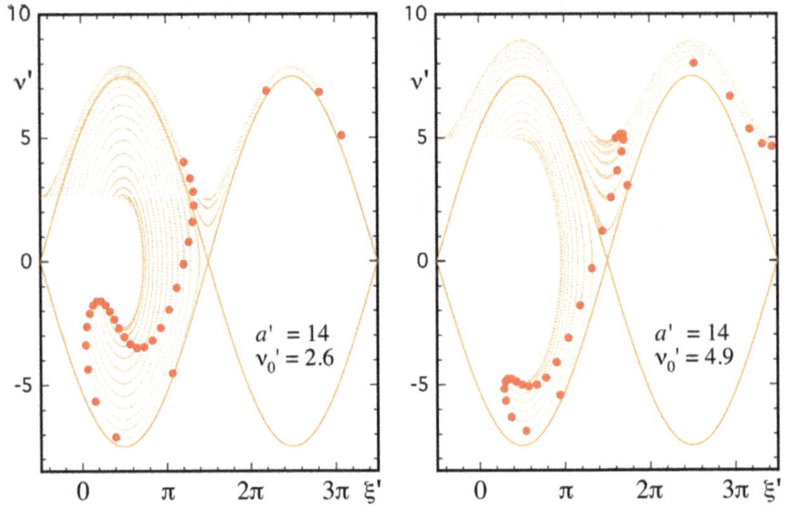

Figure 11.3. The phase space origin of the frequency pulling effect.

represented by the electrons near the stable points in phase space. From the pendulum equation, (9.53), with $\phi' = 0$, we have

$$\frac{d^2\xi'}{d\tau^2} = |a'|\cos\xi' = -|a'|\sin(\xi' - \xi_s') = -|a'|\sin\Delta\xi'; \quad \xi_s' = \frac{\pi}{2} \tag{11.5}$$

so $$\frac{d^2(\xi' - \xi_s')}{d\tau^2} = \frac{d^2\Delta\xi'}{d\tau^2} = -|a'|\sin\Delta\xi' \simeq -|a'|\Delta\xi'; \quad \Delta\xi' \ll 1. \tag{11.6}$$

The electron phase thus oscillates as $\Delta\xi' = \Delta\xi_0' \cos[\sqrt{|a'|}\,\tau]$ and the number of synchrotron oscillations in a single pass through the undulator ($\tau = 1$) equals $\sqrt{|a'|}/2\pi$. Electrons are forced to the bottom of the buckets ($\sqrt{|a'|} = \pi$) for $|a'| = \pi^2 \simeq 10$ and execute a single phase space revolution for $|a'| = 4\pi^2 \simeq 40$.

Synchrotron oscillations, when coupled with optical slippage, lead to an interesting phenomenon called the *sideband instability*. Consider two adjacent sections of an optical wave intense enough to drive one complete synchrotron oscillation, each section roughly one half of a slippage length long ($N_w\lambda/2$). Let the leading one of these sections be coincident with a given group of electrons on their downward swing at the start of the undulator; this section of the wave will be preferentially amplified due to large energy extraction from the electrons. However, by the time these electrons swing back up, they will be coincident with the trailing section of the optical wave because of slippage. This section of the wave will be attenuated due to the loss of optical energy to the coincident electrons.

Thus, the optical wave develops *spiking* in the time domain, a chaotic behavior known as the sideband instability. Its onset generally requires optical fields large enough to force at least one phase space revolution in a single pass through the

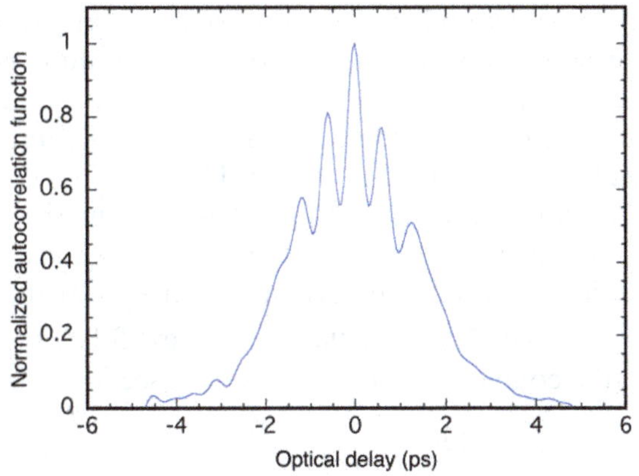

Figure 11.4. Measured autocorrelation function on the Mark III FEL at 3.2 μm showing optical spiking.

undulator, with adjacent spikes roughly separated by the slippage length $N_w\lambda$. An actual measurement of spiking is shown in figure 11.4.

The name 'sideband instability' derives from the associated perturbation in the optical spectrum. Spiking is clearly an amplitude modulation effect driven by the vertical motion of the electrons in phase space. But the accompanying horizontal motion of the electrons in their phase space orbits drives the phase of the optical wave and leads to phase modulation. One of the AM sidebands is evidently canceled by the coincident PM sideband, and the result is a single sideband on the low frequency side of the optical spectrum. Although the sideband instability leads to increased energy extraction, sideband growth is intrinsically chaotic, with sidebands eventually acquiring their own sidebands, and so on. Such spectral broadening and chaotic spiking are often undesirable for research applications using short pulses.

As noted above, the sideband instability turns on when the optical field is sufficiently large to drive at least one synchrotron oscillation, which is often the case when the fractional cavity loss of the optical resonator is sufficiently small. However, the development of spiking on multiple passes in the resonator also requires that the downward motion of the electrons in phase space at a given time τ remain successively coincident with the corresponding spike in the co-propagating optical wave at the same time τ. This is possible only if the resonator is set near its synchronous length (see section 10.4). A small degree of cavity detuning is often sufficient to extinguish the growth of the sideband instability.

11.3 Intensity dependence

In the regime of long electron bunches and laser pulses (with slippage parameter $\mu_c \lesssim 0.2$ or so), a useful and quite universal analysis of FEL saturation can be developed. In this regime, we assume that the laser spectrum is sufficiently narrow that the dynamic distortion of the gain curve at saturation does not have sufficient time to appreciably pull the optical frequency from its small-signal value within the finite duration of the macropulse. We also neglect the formation of sidebands, which assumption is justified independently. These constraints allow us to neglect all microtemporal perturbations and to base our analysis on the CW equations of motion. Short-pulse supermode effects are omitted, because laser lethargy and electron beam dispersion are negligible at saturation, as explained in section 10.5. Finally, to develop a basic quantitative understanding of saturation, we initially neglect energy spread, to be included in section 11.6.

The analytical procedure employs a straightforward numerical integration of the coupled Maxwell–Lorentz equations of motion over a single pass in the undulator, for different CW optical fields a_0' and current densities j_F, at the peak of the FEL gain curve, $\nu_0'^{\mathrm{opt}} \simeq 2.606 - 0.022 j_F + 0.00016 j_F^2$ (see text following (10.15)). For this analysis we employ the coupled equations derived in section 9.3, (9.52) and (9.53),

$$\frac{\mathrm{d}a'}{\mathrm{d}\tau} = -j_F \langle \mathrm{e}^{-\mathrm{i}\xi'} \rangle_{\xi_0'} \tag{11.7}$$

$$\frac{\mathrm{d}\nu'}{\mathrm{d}\tau} = |a'| \cos(\xi' + \phi'), \tag{11.8}$$

where $j_F = \langle j_L |f_p|^2 R_F \rangle$ and $|a'|^2 = |a_p|^2 \langle R_F^2 |f_p|^2 \rangle$. Energy conservation, (9.49), is written

$$-2 j_F \langle \Delta\nu \rangle_{\xi_0} = \Delta |a'|^2 \qquad \text{or} \qquad -2\frac{j_L}{R_F}\langle \Delta\nu \rangle_{\xi_0} = \Delta |a_p|^2. \tag{11.9}$$

where $\Delta\nu' = \Delta\nu$. Upon solving (11.7) and (11.8) numerically at peak $\nu_0'^{\mathrm{opt}}$, we can approximate the results by the following expression:

$$G_{\mathrm{sat}} \equiv \frac{\Delta |a'|^2}{|a'|^2} = \frac{G'(j_F)}{\left[1 + \left(\frac{|a'|}{4\pi}\right)^2\right]^{\sqrt{\pi}}}, \tag{11.10}$$

where $G'(j_F)$ is given in (10.15). The coefficients '4π' and '$\sqrt{\pi}$' are *mnemonic*; numerically, we find

$$`4\pi' = 13.60 - 0.257 j_F + 0.00942 j_F^2 - 0.000150 j_F^3 \tag{11.11}$$

$$`\sqrt{\pi}' = 1.800 - 0.0114 j_F + 0.000864 j_F^2 - 0.0000159 j_F^3. \tag{11.12}$$

For $j_F \leqslant 22$ these expressions convey an accuracy of $|\Delta G_{\mathrm{sat}}|/G_{\mathrm{sat}} < 3.5\%(7\%)$ for $|a'| \leqslant 19(30)$. Nevertheless, considerable accuracy is retained in analyses using the mnemonic values 4π and $\sqrt{\pi}$. We see that the intensity dependence of FEL saturation, obtained here by rigorous numerical integration of the equations of motion, manifestly does not correspond to the saturation mechanism of homogeneously broadened atomic lasers, $G \sim [1 + I/I_{\mathrm{sat}}]^{-1}$. This has important implications for the analysis of optical resonators, as we discuss below.

11.4 Analysis of optical resonators

Consider a generic FEL optical resonator with fractional cavity loss δ_c and output coupling δ_{oc}, illustrated in figure 11.5. In the small signal regime, the net gain in a single round trip is

$$1 + G_{net} = (1 + G_{ss})(1 - \delta_c) > 1, \tag{11.13}$$

where G_{ss} is the small-signal, single-pass laser gain; see section 10.5.

For steady state oscillation at saturation, we have

$$1 = (1 + G_{sat})(1 - \delta_c); \qquad 1 + G_{sat} = \frac{1}{1 - \delta_c}; \qquad G_{sat} = \frac{\delta_c}{1 - \delta_c}. \tag{11.14}$$

The output power is thus proportional to

$$|a'|^2_{out} = |a'|^2(1 + G_{sat})\delta_{oc} \tag{11.15}$$

$$= |a'|^2 \frac{\delta_c}{1 - \delta_c}\left(\frac{\delta_{oc}}{\delta_c}\right) \tag{11.16}$$

$$= |a'|^2 G_{sat}\eta_{oc}; \tag{11.17}$$

$$\text{or} \quad |a'|^2_{out} = |a'|^2\, \eta_{oc} \frac{G'}{\left[1 + \left(\frac{|a'|}{4\pi}\right)^2\right]^{\sqrt{\pi}}}, \tag{11.18}$$

where we define the *output coupling efficiency* $\eta_{oc} \equiv \frac{\delta_{oc}}{\delta_c}$ as the fraction of total cavity loss that appears as usable output coupling. This parameter is typically known by design. From the known cavity losses δ_c we can calculate the steady-state saturated gain G_{sat} from (11.14), then calculate $|a'|^2$ at saturation from (11.10) and finally calculate $|a'|^2_{out}$ from (11.17).

Optimum output coupling. What is the optimum cavity loss δ_c for a known value of the small-signal gain G'? Physically, we might expect there to be some optimum value: if δ_c is too large then no power builds up, and if δ_c goes to zero (including δ_{oc}!) then no power gets out.

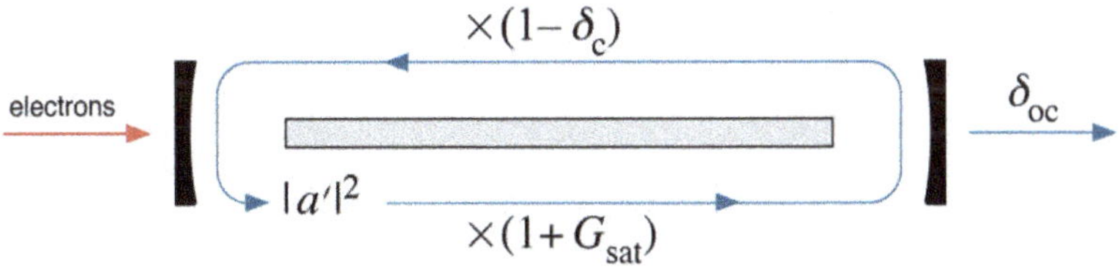

Figure 11.5. Fractional gain and loss factors in an optical resonator.

Since equation (11.18) gives $|a'|^2_{out}$ as a function of $|a'|^2$, we can maximize $|a'|^2_{out}$ versus $|a'|^2$ by differentiation. Using mnemonic values for the coefficients, the result is

$$|a'|^2 = |a'|^2_{opt} = \frac{(4\pi)^2}{\sqrt{\pi} - 1} \quad \rightarrow \quad |a'|_{opt} = 14.3. \tag{11.19}$$

This value of the optimum intracavity power is completely independent of the design of the optical resonator! For this value of $|a'|_{opt}$, the electrons undergo 0.6 phase space revolutions (see figure 11.3), so the assumption that sidebands do not turn on is a sensible one. It is perhaps not surprising that optimum energy extraction should correspond to electrons that have descended to the bottom of the phase space buckets. The associated saturated gain is

$$G_{sat} = \frac{G'}{\left[1 + \dfrac{1}{\sqrt{\pi} - 1}\right]^{\sqrt{\pi}}} = \frac{G'}{4.36}, \tag{11.20}$$

and the optimum cavity losses from (11.14) are

$$\delta_c^{opt} = \frac{G_{sat}}{1 + G_{sat}} = \frac{G'}{4.36 + G'}. \tag{11.21}$$

Finally, the optimum output power from (11.17) is proportional to

$$|a'|^2_{out} = |a'|^2_{opt} G_{sat} \eta_{oc} = 46.9\, G' \eta_{oc}. \tag{11.22}$$

Of course, greater accuracy is obtained in this analysis using the previous expansions for the coefficients '4π' and '$\sqrt{\pi}$' from (11.11) and (11.12).

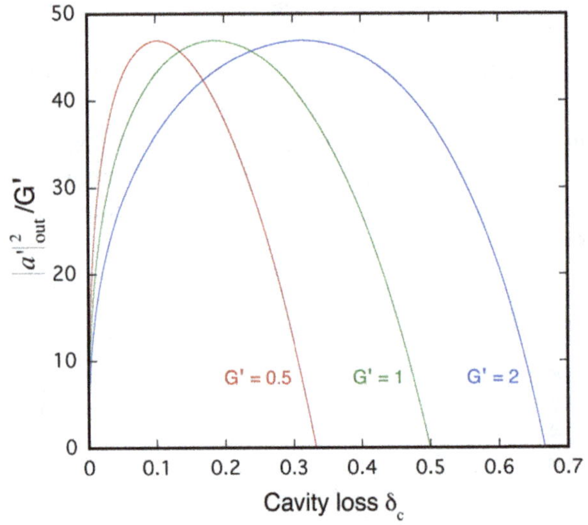

Figure 11.6. Outcoupled optical power as a function of cavity loss.

These results exemplify a fundamental difference in the saturation mechanism in FELs as opposed to conventional lasers, which distinction can be traced back to the appearance of the exponent $\sqrt{\pi}$ in the denominator of (11.10): in conventional lasers, the corresponding exponent does not exceed unity and it is not possible to optimize the total cavity loss δ_c. Instead, it is only possible to optimize the outcoupled cavity loss δ_{oc} for a specified value of the non-outcoupled cavity loss δ_{fixed}, where $\delta_c = \delta_{oc} + \delta_{fixed}$.

Optical power. The conversion of optical power between $|a'|^2$ and $|a_p|^2$ was calculated in (10.13),

$$|a'|^2 = 2q_E|a_p|^2, \tag{11.23}$$

where q_E is defined in (10.11). The optical power in the TEM_{00} mode (with the subscript indicating *optical*) is

$$P_{op} = \iint \mathrm{d}x\,\mathrm{d}y \frac{c}{4\pi}|\hat{E}|^2 \tag{11.24}$$

$$= \frac{c}{4\pi}\pi w_0^2 \iint \frac{\mathrm{d}x\,\mathrm{d}y}{\pi w_0^2}|c_p|^2 u_p^* u_p \tag{11.25}$$

$$= \frac{c}{4\pi}\pi w_0^2\,|c_p|^2; \tag{11.26}$$

$$\text{or}\quad P_{op}[\text{watts}] = 10^{-7}\left[\frac{\text{joules}}{\text{erg}}\right] \times \frac{\gamma^4 m^2 c^5}{32\,\pi^3 e^2 N_w^4 \lambda_w^2 \hat{K}_f^2} \cdot \frac{\pi w_0^2}{2}|a_p|^2, \tag{11.27}$$

where all physical quantities on the rhs are in CGS units and we converted from c_p to a_p using (8.14).

If we calculate $|a'|^2_{out}$ as a function of δ_c (with $\eta_{oc} = 1$), we find that the factor of '46.9' in (11.22) varies as

$$\frac{|a'|^2_{out}}{G'} = (4\pi)^2\left(\left[\frac{G'(1-\delta_c)}{\delta_c}\right]^{1/\sqrt{\pi}} - 1\right)\frac{\delta_c}{G'(1-\delta_c)}. \tag{11.28}$$

These curves are shown in figure 11.6 for $G' = 0.5$; 1; 2, and are quite broad; the range of δ_c for which $|a'|^2_{out}$ exceeds 90% of its maximum value is on the order of δ_c itself. Thus, the *net* round-trip, small-signal gain can be substantially increased by decreasing the cavity loss δ_c at little expense to the outcoupled power at saturation.

Example. What are the optimum cavity losses and output power for the example in section 10.5? Use both the mnemonic values of 4π and $\sqrt{\pi}$ and their numerical expansions. Assume $q_\gamma = 1$ and $\eta_{oc} = 1$.

Solution. In that example we calculated $j_F = 7.792$, for which

$$G'(j_F) = 1.355 \tag{11.29}$$

from (10.15). Using mnemonic results from (11.21) and (11.22), we find $\delta_c^{opt} = 23.7\%$ and $|a'|_{out}^2 = 63.5$. With the previous value of $q_E = 0.700$ we have

$$|a_p|_{out}^2 = \frac{|a'|_{out}^2}{2q_E} = 45.4. \tag{11.30}$$

Substituting this value for $|a_p|_{out}^2$ into (11.27), together with the parameters listed in the original example in section 9.1, yields an outcoupled power of $P_{out} = 8.81$ MW.

If we now use the full expansions for '4π' and '$\sqrt{\pi}$' from (11.11) and (11.12), we calculate '4π' = 12.10 and '$\sqrt{\pi}$' = 1.756. From (11.19) we have

$$|a'|^2 = |a'|_{opt}^2 = \frac{(4\pi)^2}{\sqrt{\pi} - 1} = \frac{(12.10)^2}{1.756 - 1} \quad \rightarrow \quad |a'|_{opt} = 13.92, \tag{11.31}$$

and the corresponding saturated gain from (11.20) is

$$G_{sat} = \frac{G'}{\left[1 + \dfrac{1}{\sqrt{\pi} - 1}\right]^{\sqrt{\pi}}} = \frac{G'}{\left[1 + \dfrac{1}{1.756 - 1}\right]^{1.756}} = \frac{G'}{4.39}. \tag{11.32}$$

The optimum cavity losses and output power from (11.14) and (11.17) are then

$$\delta_c^{opt} = \frac{G_{sat}}{1 + G_{sat}} = \frac{G'}{4.39 + G'} = 23.6\%, \tag{11.33}$$

$$|a'|_{out}^2 = |a'|_{opt}^2 G_{sat} = 44.1\ G' = 59.8. \tag{11.34}$$

The resulting output power is $P_{out} = 8.3$ MW. Measured outcoupled optical powers from the Mark III FEL are actually of this magnitude—divided into four separate pulses, one from each surface of the intracavity Brewster plate output coupler.

11.5 Extraction efficiency

Into the second of (11.9) for energy conservation,

$$-2\frac{j_L}{R_F}\langle \Delta\nu \rangle_{\xi 0} = \Delta |a_p|^2, \tag{11.35}$$

substitute the expressions for $\Delta\nu$, j_L, R_F and $\Delta|a_p|^2$ in CGS units from (7.36, 8.15, 8.32, 8.38, 11.27):

$$-2 \cdot \frac{8\pi^2 e^2 N_w^3 \lambda_w^2 \hat{K}_f^2}{\gamma^3 mc^2} \frac{(I/e)_{MKS}}{c\ A_e} \cdot \frac{A_e}{\pi w_0^2} \cdot 4\pi N_w \left\langle \frac{\Delta\gamma}{\gamma} \right\rangle$$
$$= \Delta P_{op} \cdot \frac{2}{\pi w_0^2} \cdot \frac{32\ \pi^3 e^2 N_w^4 \lambda_w^2 \hat{K}_f^2}{\gamma^4\ m^2 c^5}. \tag{11.36}$$

This simplifies to

$$\Delta P_{\text{op}} = (I/e)_{\text{MKS}} \langle \Delta(\gamma mc^2) \rangle \tag{11.37}$$

$$\Delta P_{\text{op}} = I_{\text{MKS}} \frac{\gamma mc^2}{e_{\text{MKS}}} \left\langle \frac{\Delta\gamma}{\gamma} \right\rangle \tag{11.38}$$

$$\Delta P_{\text{op}} = I_{\text{MKS}} V_{\text{MKS}} \left\langle \frac{\Delta\gamma}{\gamma} \right\rangle, \tag{11.39}$$

where ΔP_{op} is now in watts. If the laser is oscillating in steady state, the power extracted from the electron beam on a single pass must leave the cavity on that pass, i.e. $\Delta P_{\text{op}} = P_{\text{out}}$. Thus, we have a quick way to calculate P_{out}, *if* we know the *extraction efficiency* $\langle \Delta\gamma/\gamma \rangle$. To obtain this, use the first of (11.9) for energy conservation,

$$2j_F \langle \Delta\nu \rangle_{\xi 0} = \Delta |a'|^2 = G_{\text{sat}} |a'|^2 \qquad \text{(in general)} \tag{11.40}$$

$$= \frac{G'}{4.36}(14.3)^2 \qquad \text{(optimum cavity loss; mnemonic).} \tag{11.41}$$

For optimum cavity losses we thus have, using mnemonic coefficients,

$$4\pi N_w \left\langle \frac{\Delta\gamma}{\gamma} \right\rangle = \langle \Delta\nu \rangle = 46.9 \frac{G'}{2j_F}, \tag{11.42}$$

$$\text{or} \quad \left\langle \frac{\Delta\gamma}{\gamma} \right\rangle = 1.866 \frac{G'}{j_F N_w}; \quad \left(\rightarrow \frac{1}{4N_w} \text{ as } j_F \rightarrow 0 \right). \tag{11.43}$$

As an illustration, in the previous example we had $j_F = 7.792$ and $G' = 1.355$, so $\langle \Delta\gamma/\gamma \rangle = \frac{0.3245}{N_W}$. The output power from (11.39) is $P_{\text{out}} = I_{\text{MKS}} V_{\text{MKS}} \langle \Delta\gamma/\gamma \rangle =$ (30 A)(42.511 MeV)(0.006904) = 8.80 MW, as calculated previously using mnemonic coefficients.

More generally, if j_F and δ_c are arbitrary, so that the cavity losses are not optimized, then the procedure for calculating the extraction efficiency is to find G_{sat} from (11.14) and $|a'|^2$ from (11.10). Then the decrease in phase velocity is $\langle \Delta\nu \rangle = G_{\text{sat}} |a'|^2 / 2j_F$, the extraction efficiency is $\langle \Delta\gamma/\gamma \rangle = \langle \Delta\nu \rangle / 4\pi N_w$, and the output power is $P_{\text{out}} = \eta_{\text{oc}} I_{\text{beam}} V_{\text{beam}} \langle \Delta\gamma/\gamma \rangle$.

The extraction efficiency is inversely proportional to the number of undulator periods N_w (11.43). Therefore, an FEL with fewer undulator periods will yield larger optical power at saturation (11.39), all other things being equal. Of course, the laser gain also decreases with fewer periods, so there is a trade-off in the design of the FEL. However, we do not have to change the design of the FEL to observe an

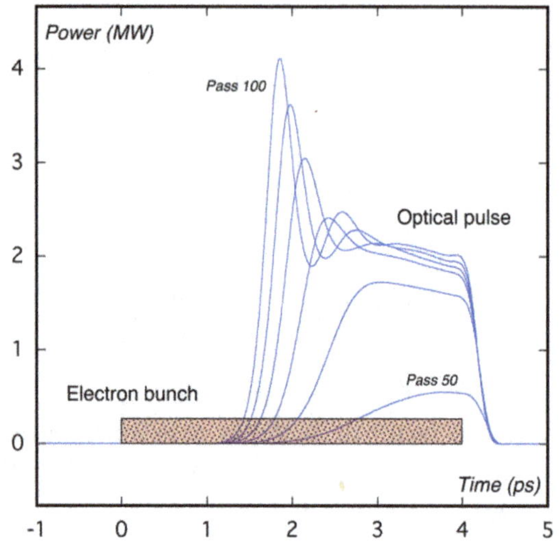

Figure 11.7. Numerical simulation of laser oscillation near the synchronous cavity length.

enhancement in extraction efficiency: there is an interesting phenomenon related to this effect that appears in numerical FEL pulse propagation simulations and provides a nice illustration of the physics involved.

Figure 11.7 is the result of a numerical simulation showing the formation of a saturated optical pulse at 3.2 μm in the Mark III FEL, operating near the synchronous cavity length. The FEL is driven by a train of rectangular electron bunches located between $t = 0$ ps and $t = 4$ ps in the numerical window, with the leading edge on the left. The figure shows the relative overlap between the two pulses at the start of the undulator at ten-pass intervals in the resonator. We see two interesting effects. First, by the time the optical pulse enters the large signal regime around pass 50, laser lethargy in the small signal regime has pushed the optical pulse towards the trailing edge of the electron bunch. The laser evidently saturates before the optical pulse walks off entirely, at which point lethargy is 'frozen' and the leading edge moves forward again under the combined action of the laser interaction and the finite (albeit small) cavity detuning. The second thing we observe is the abrupt formation of a narrow spike at the leading edge of the optical pulse with more than twice the peak power of the trailing section of the pulse.

What is the origin of this spike? The answer is the $1/N_w$ dependence of the FEL extraction efficiency. As the laser evolves more deeply into saturation, most of the electrons in the bunch (between ~1.5–4 ps on the latter passes) overlap and interact with the optical pulse for the entire length of the undulator, the full N_w periods. But the electrons at the leading edge of the bunch are not overlapped at all when the pulses first enter the undulator. Electrons positioned within one slippage length in advance of the sharp, leading edge of the optical pulse will not participate in the FEL interaction until the optical pulse slips past them some distance along the undulator. These electrons interact with the optical field over a reduced number of undulator periods $N_w^{\mathrm{eff}} < N_w$. As a result, they experience greater extraction efficiency, and dump their energy at the leading edge of the optical pulse in an increasingly tall and narrow spike roughly one slippage length wide. This spike often serves as a seed for the formation of sidebands: the smaller ripples following the leading spike in the optical pulse are actually the onset of the sideband instability.

11.6 Incorporation of energy spread

To include energy spread in the analysis of saturation, we perform a numerical integration of the CW coupled equations, (9.52) and (9.53),

$$\frac{\mathrm{d}a'}{\mathrm{d}\tau} = -j_F \langle \mathrm{e}^{-\mathrm{i}\xi'} \rangle_{\xi_0', \nu_0'} \tag{11.44}$$

$$\frac{\mathrm{d}\nu'}{\mathrm{d}\tau} = |a'| \cos(\xi' + \phi'), \tag{11.45}$$

but we include a second dimension in the initial phase space distribution to impose a normalized Gaussian energy spread in ν_0', (10.22), on the initially uniform distribution in phase ξ_0'. The same physical constraints—absence of frequency pulling, sideband formation, and short-pulse perturbations—are assumed to apply here as in section 11.3. For small optical fields, the small-signal gain obtained from the numerical integration of (11.44) and (11.45) agrees with the gain obtained from a numerical integration of the weak field solution, (10.27), with an accuracy of $|\Delta G|/G < 0.36\%$ over the full range $\sigma \leqslant 4$ and $j_F \leqslant 22$. Integration of the coupled equations for large optical fields satisfies energy conservation to within roughly twice this error.

To achieve useful approximations to the solution for large optical fields over the full range of σ and j_F, we first derive an improved approximation to the inhomogeneous gain reduction factor q_γ in the calculation of the small-signal gain, (10.28). The revised calculation allows the two numerical coefficients in (10.24) to include a dependence on j_F, yielding a new gain reduction factor q_γ' given by

$$q_\gamma' \simeq \frac{1}{\left[1 + \left(\dfrac{\sigma}{A(j_F)}\right)^2\right]^{B(j_F)}}, \tag{11.46}$$

where

$$A(j_F) = 3.27 + 0.0183\, j_F - 0.000169\, j_F^2 \tag{11.47}$$

$$B(j_F) = 1.80 + 0.0138\, j_F - 0.000160\, j_F^2. \tag{11.48}$$

The use of this gain reduction factor in the calculation of the small-signal gain,

$$G_\gamma' = G'(j_F q_\gamma'), \tag{11.49}$$

where $G'(...)$ is the three-term expansion defined by (10.15), yields an accuracy of $|\Delta G'_\gamma|/G'_\gamma < 1\%$ for $\sigma \leqslant 4$ and $j_F \leqslant 22$.

The intensity dependence of the saturated gain G_{sat} in the presence of energy spread is then obtained by fitting each of the coefficients '4π' and '$\sqrt{\pi}$' in (11.10) to a series expansion in products $j_F^n \sigma^m$. Upon solving (11.44) and (11.45) numerically at peak $\nu_0'^{\text{opt}}$, we can approximate the results by the following expression:

$$G'_{\text{sat}} \equiv \frac{\Delta|a'|^2}{|a'|^2} = \frac{G'_\gamma}{\left[1 + \left(\frac{|a'|}{\alpha_\gamma}\right)^2\right]^{\beta_\gamma}}, \tag{11.50}$$

where the parameters α_γ and β_γ are given by the power series expansions

$$\alpha_\gamma = \sum_{n=0}^{3}\sum_{m=0}^{4} \alpha_{nm} j_F^n \sigma^m; \quad \beta_\gamma = \sum_{n=0}^{3}\sum_{m=0}^{4} \beta_{nm} j_F^n \sigma^m, \tag{11.51}$$

and the numerical coefficients α_{nm} and β_{nm} are the corresponding elements of the following coefficient matrices:

$$\alpha_{nm} = \begin{pmatrix} 13.60 & -0.277 & 2.09 & -0.367 & 0.0198 \\ -0.257 & 0.00833 & -0.0465 & 0.0170 & -0.00164 \\ 0.00942 & 0.000107 & 0.000272 & -0.000307 & 0.0000409 \\ -0.000150 & -3.31\times10^{-6} & 5.87\times10^{-6} & 2.08\times10^{-6} & -4.44\times10^{-7} \end{pmatrix}, \tag{11.52}$$

$$\beta_{nm} = \begin{pmatrix} 1.800 & -0.0287 & 0.177 & -0.0463 & 0.00352 \\ -0.0114 & 0.00150 & -0.00924 & 0.00330 & -0.000321 \\ 0.000864 & -0.0000115 & 0.000157 & -0.0000798 & 9.13\times10^{-6} \\ -0.0000159 & 8.44\times10^{-8} & -1.36\times10^{-6} & 1.02\times10^{-6} & -1.32\times10^{-7} \end{pmatrix}. \tag{11.53}$$

For $\sigma \leqslant 4$ and $j_F \leqslant 22$ these coefficients convey an accuracy of $|\Delta G'_{\text{sat}}|/G'_{\text{sat}} < 3.5\%(8\%)$ for $|a'| \leqslant 18(29)$.

Numerical values for G'_γ, α_γ and β_γ obtained from the above expressions can be used directly in the analysis of optical resonators developed in section 11.4. For example, the optimum intracavity power, saturated gain, cavity loss and output power from (11.19)–(11.22) are

$$|a'|^2_{\text{opt}} = \frac{(\alpha_\gamma)^2}{\beta_\gamma - 1}; \quad G'_{\text{sat}} = \frac{G'_\gamma}{\left[1 + \frac{1}{\beta_\gamma - 1}\right]^{\beta_\gamma}}; \quad \delta_c^{\text{opt}} = \frac{G'_{\text{sat}}}{1 + G'_{\text{sat}}}; \quad |a'|^2_{\text{out}} = |a'|^2_{\text{opt}} G'_{\text{sat}} \eta_{\text{oc}}. \tag{11.54}$$

The application of these results will be illustrated in section 12.2, where the effects of energy spread are of particular significance for lasing on higher harmonics.

Numerical approximations notwithstanding, the above analysis reveals some interesting physics about the effects of energy spread on the saturation mechanism in FELs. First, examine the dependence of the coefficients α_γ and β_γ on the energy spread σ. These dependencies are plotted in figure 11.8.

The coefficient β_γ increases only slightly with energy spread, from ~1.8 to 2.4 over the range of σ shown, but the coefficient α_γ roughly doubles over the same range. Since α_γ^2 plays the role of a 'saturation intensity', the increase in α_γ with energy spread leads to larger saturated powers than would otherwise ensue if α_γ and β_γ were independent of σ. This behavior counteracts the detrimental effect of inhomogeneous gain reduction in the small-signal regime. The corresponding effect on the output power $|a'|^2_{\rm out}$ is illustrated in figure 11.9 for a current of $j_F = 8$. These curves

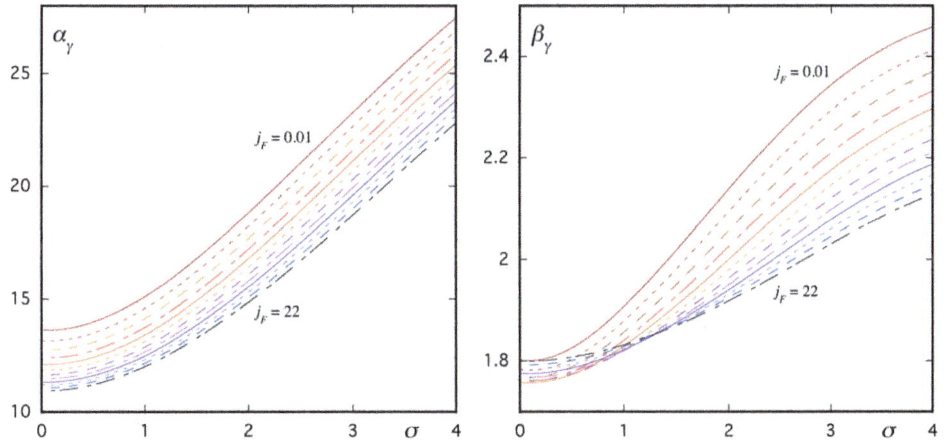

Figure 11.8. FEL saturation coefficients α_γ and β_γ versus energy spread σ; $j_F = 0.01, 2, 4, 6, \ldots, 22$.

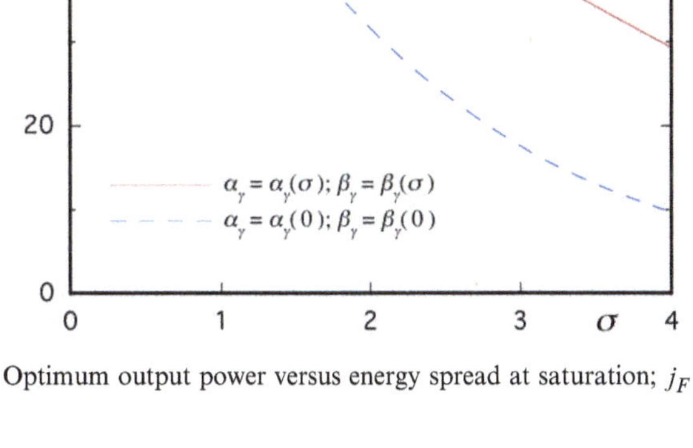

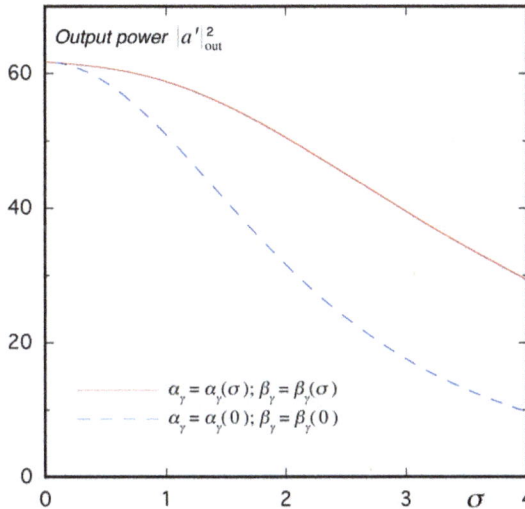

Figure 11.9. Optimum output power versus energy spread at saturation; $j_F = 8$; $\eta_{\rm oc} = 1$.

show the optimum output power calculated from (11.54) assuming the full variation of G_γ' with energy spread according to (11.49)—but the dashed curve is the optimum output power calculated with constant coefficients $\alpha_\gamma(\sigma = 0)$; $\beta_\gamma(\sigma = 0)$, while the solid curve assumes the full σ-dependence $\alpha_\gamma(\sigma)$; $\beta_\gamma(\sigma)$ given by (11.51). We see that the σ-dependence of the coefficients yields a significant increase in power at all values of the energy spread. Equation (11.54) reveals that the optimum output power $|a'|^2_{\text{out}}$ with fixed coefficients is strictly proportional to the small-signal gain G_γ', i.e. the output power is afflicted by the full effect of inhomogeneous gain reduction. The enhancement of output power that actually occurs for variable coefficients motivates the aphorism that 'the FEL interaction is more accommodating of energy spreads at saturation'.

The energy acceptance of the FEL can be equivalently defined in both the small signal and large signal regimes. In the small signal regime, the form of the small signal gain curve requires that the $1/e$ energy half-width satisfy $\Delta\nu_0 \leqslant 2.6$, so that all the electrons contribute a positive gain. In the optimized FEL interaction at saturation, the behavior of the electron orbits in phase space similarly requires $\Delta\nu_0 \leqslant 2.6$, so that all the electrons undergo half a phase space revolution in the *downward* direction. Since $\Delta\nu_0 = \sqrt{2}\sigma$, we require in either case that $\sigma \lesssim 2.6/\sqrt{2} \simeq 2$, or $(\delta\gamma/\gamma)_{1/e} \lesssim 1/2N_w$ (10.25). This is the *energy acceptance* of the FEL and yields a fundamental constraint on the quality of the energy spread in the electron beam.

Chapter 12

Harmonic lasing

12.1 Small-signal gain

Equation (5.21) indicates that on-axis spontaneous emission in a plane-polarized undulator occurs at all odd harmonics $f = 1, 3, 5, \ldots$ of the fundamental frequency. The microscopic origin of this emission is illustrated in figure 4.1. Typically, of course, laser oscillation occurs at the fundamental frequency $f = 1$. But if one of the other spontaneous harmonics is synchronous with the electron bunches on successive passes in the resonator, then amplification of that harmonic due to stimulated emission can occur by exactly the same process as described in chapter 1. The FEL interaction in this case is described by the Maxwell–Lorentz equations for the specific harmonic number f. The synchronous cavity lengths for the individual harmonics can easily be separated in an RF-linac FEL by inserting a small degree of optical dispersion into the cavity.

To analyze FEL operation on a particular harmonic f, return to the FEL ansatz in section 9.3. For operation on harmonic f, (9.44) and (9.45) are reduced under the assumed approximations to

$$\frac{\mathrm{d}a'}{\mathrm{d}\tau} = -j_F \langle \mathrm{e}^{-\mathrm{i}(f\xi - \measuredangle f_p)} \rangle_{\xi_0, \nu_0} \tag{12.1}$$

$$\frac{\mathrm{d}\nu}{\mathrm{d}\tau} = |a'| \cos(f\xi - \measuredangle f_p + \phi'), \tag{12.2}$$

where $j_F \equiv \langle j_L |f_p|^2 R_F \rangle_{\text{und}}$ and $|a'|^2 \equiv |a_p|^2 \langle R_F^2 \, |f_p|^2 \rangle_{\text{und}}$. Make the following change of variables:

$$a_h' = f a' \tag{12.3}$$

$$\xi' = f\xi - \measuredangle f_p \tag{12.4}$$

$$\nu' \equiv \frac{\mathrm{d}\xi'}{\mathrm{d}\tau} = f\nu - \frac{\mathrm{d}}{\mathrm{d}\tau}\measuredangle f_p . \tag{12.5}$$

If $\measuredangle f_p$ again has only a first-order τ-dependence (section 9.4), then the equations of motion, (12.1) and (12.2), assume the standard form

$$\frac{\mathrm{d}a_h'}{\mathrm{d}\tau} = -fj_F \langle \mathrm{e}^{-\mathrm{i}\xi'} \rangle_{\xi_0', \nu_0'} \tag{12.6}$$

$$\frac{\mathrm{d}\nu'}{\mathrm{d}\tau} = |a_h'| \cos(\xi' + \phi_h'). \tag{12.7}$$

We see that harmonic lasing is readily accommodated by our earlier analyses if we simply replace $j_F \to fj_F$. For example, the analytic small-signal gain at the harmonic f (for small gain) is

$$G = 0.135(fj_F). \tag{12.8}$$

Although the harmonic gain formally appears to be greater than the fundamental gain by a factor of f, there is other dependence on the harmonic number in the expression for j_F and other gain reduction factors, and the harmonic gain rarely exceeds the gain at the fundamental in real systems.

Following the prescription summarized in section 10.5, the small-signal harmonic gain is calculated as follows:

1) $$j_F = \frac{8\pi^2 e^2 N_w^3 \lambda_w^2 \hat{K}_f^2}{\gamma^3 mc^2} \frac{(I/e)_{\mathrm{MKS}}}{c} \frac{q_E}{A_{\mathrm{opt}}}; \quad \text{(see (10.11) for } q_E) \tag{12.9}$$

2) $$G_\gamma = G'(fj_F q_\gamma); \quad \text{(see (10.24) for } q_\gamma \text{ and (10.15) for } G') \tag{12.10}$$

3) $$G_{\mathrm{ss}} = G_\gamma \, \mathcal{G}(\rho, \mu_c); \quad \text{(see (10.34) for } \mathcal{G}(\rho, \mu_c)). \tag{12.11}$$

The undulator parameter $\hat{K}_f^2$ is given in (6.20), with η given in (4.21). The gain reduction factor q_E due to beam overlap is given by (10.11). However, the optical mode radius w_0 in that expression and the optical mode area A_{opt} in (12.9) are those of the harmonic mode: if the same optical resonator is employed for both the fundamental and harmonic (as is typically the case), then the Rayleigh range is common to both and the area of the harmonic mode is reduced by a factor of $\lambda_h/\lambda_1 = 1/f$. This effect helps to increase the harmonic gain to practical values for laser operation.

Short-pulse effects can be included via the factor $\mathcal{G}(\rho, \mu_c)$ as per the prescription summarized in section 10.4, because the slippage parameter has the same numerical value for all harmonics.

However, inhomogeneous gain reduction due to energy spread is exacerbated at the higher harmonics, because the fractional spectral widths of the spontaneous spectrum and gain curve are decreased by a factor of $1/f$ due to the scaling prescribed in (12.5). As a result, the fixed fractional energy spread covers a proportionally larger fraction of the gain curve. Inhomogeneous gain reduction can be included using (10.24) and (10.25), but due to the narrowing of the optical spectrum, the effective energy spread employed in those equations needs to be increased by a factor of f.

Example. Calculate the small-signal gain at the fundamental and third harmonic wavelengths in a Mark III-class FEL at a cavity length detuning of $\delta L_c^p/3$, assuming the following parameters:

1) Peak current $I = 30$ A
2) Fundamental wavelength $\lambda_1 = 4.8$ µm
3) Undulator parameter $\hat{K}^2 = 1.20$
4) Undulator period $\lambda_w = 2.3$ cm
5) Number of periods $N_w = 47$
6) Rayleigh range $z_R = 53$ cm
7) $\epsilon_x^n = 8\pi$ mm·mrad; $\beta_x = 53$ cm (matched)
8) $\epsilon_y^n = 4\pi$ mm·mrad; $\beta_y = 24$ cm (matched)
9) 1/e fractional energy spread 0.3%
10) Full-width bunch length $2\sigma_z/c = 2$ ps

Solution. We calculate the following parameters at the fundamental and third harmonic:

Parameter	$f=1$	$f=3$
$\hat{K}_f^2$	0.860	0.106
γ	72.6	72.6
w_0	900 µm	520 µm
$A_{\rm opt}$	0.0127 cm^2	0.00424 cm^2
w_x	242 µm	242 µm
w_y	115 µm	115 µm
ζ_w	1.020	1.020
q_E	0.718	0.616
j_F	9.698	3.072
$\sigma_{\rm eff} = f\sigma$	0.626	1.879
$q_\gamma(\sigma_{\rm eff})$	0.937	0.598
G_γ	1.641	0.894
μ_c	0.753	0.753
$\mathcal{G}(\rho_p/3, \mu_c)$	0.559	0.559
G_{ss}	91.7%	50.0%

The gain at the third harmonic is lower than at the fundamental, but the harmonic gain is still quite robust and harmonic lasing is in fact a useful way to extend the wavelength range of the laser.

12.2 Saturation and output power

The optical resonator and saturation analyses of sections 11.4 and 11.6 carry over for harmonic lasing. The only remaining formal substitution we have to make, as prescribed by (12.3) and (10.13), is to set

$$|a'|^2 \rightarrow |a_h'|^2 = f^2 |a'|^2 \tag{12.12}$$

$$= f^2\, 2q_E |a_p|^2, \tag{12.13}$$

where a_p is the dimensionless version of the complex mode coefficient c_p of the pth transverse mode in the expansion of the optical field ((8.7) and (8.14)) and is used in (11.27) to calculate the power in watts. The output power can also be calculated directly from the extraction efficiency. It is left as a (worthy!) exercise for the student to show that the optimum output power is given by $P_{out} = \eta_{oc}\, I_{MKS}\, V_{MKS} \langle \Delta\gamma/\gamma \rangle$, where the extraction efficiency for the fth harmonic (based on the analysis of section 11.6) is

$$\left\langle \frac{\Delta\gamma}{\gamma} \right\rangle = \frac{1}{2N_w}\left[\frac{G_\gamma'}{4\pi f^2 j_F}\right]\left[\frac{\alpha_\gamma^2}{\beta_\gamma^{\beta_\gamma}(\beta_\gamma - 1)^{1-\beta_\gamma}}\right], \tag{12.14}$$

and $G_\gamma' = G'(fj_F\, q_\gamma')$ (from (11.49), with $j_F \rightarrow fj_F$). This is clearly the easiest way to calculate P_{out}.

Example. What are the optimum cavity loss, net round trip gain and optimum output power at saturation for each of the fundamental and third harmonic wavelengths in the example in the preceding section? Take $\eta_{oc} = 1$ and use the analysis of section 11.6.

Solution.

1) The first step is to calculate the small-signal gain G_γ', which differs slightly from the gain G_γ calculated in the previous example. We construct the following table of parameters:

Parameter	$f = 1$	$f = 3$
j_F	9.698	3.072
$\sigma_{eff} = f\sigma$	0.626	1.879
$A(fj_F)$	3.432	3.424
$B(fj_F)$	1.919	1.914
$q_\gamma'(\sigma_{eff})$	0.939	0.604
$G_\gamma'(fj_F\, q_\gamma')$	1.646	0.906

2) The next step is to calculate the saturation parameters α_γ and β_γ from (11.51). These calculations can be expedited on a computer, and we obtain the following (remembering to set $j_F \to fj_F$ and $\sigma \to \sigma_{\text{eff}}$):

Parameter	$f=1$	$f=3$
α_γ	12.37	16.07
β_γ	1.782	1.965

3) At this point we have everything we need to calculate P_{out} using (12.14), and we obtain $P_1^{\text{out}} = 7.2$ MW and $P_3^{\text{out}} = 2.0$ MW.
4) We continue with the calculation of the optimum cavity loss and net round trip gain. The basic resonator and laser parameters for optimum cavity loss are given by (11.54), with $a' \to a'_h$ formally, and the results are as follows:

Parameter	$f=1$	$f=3$
$\lvert a'_h \rvert_{\text{opt}}$	13.99	16.36
G'_{sat}	0.379	0.224
δ_c^{opt}	27.5%	18.3%
$\lvert a'_h \rvert^2_{\text{out}}$	74.2	60.0

The analysis of saturation employs the gain factor G'_γ. However, if we are interested in the net round trip gain G_{net} in the small signal regime in which we are actually operating, then, from (11.13), we need to consider the actual small signal gain G_{ss}, which includes the short-pulse gain reduction factor. The values for G_{ss} obtained in the previous example are adequate for this calculation, and we obtain

Parameter	$f=1$	$f=3$
G_{ss}	91.7%	50.0%
δ_c^{opt}	27.5%	18.3%
G_{net}	39.0%	22.6%

5) To confirm the values for the output power obtained in step 3, we convert the values for $|a_h'|^2_{\text{out}}$ from step 4 to the corresponding values for $|a_p|^2_{\text{out}}$ and P_{out} using (12.13) and (11.27). These calculations indeed confirm:

Parameter	$f=1$	$f=3$
$\|a_p\|^2_{\text{out}}$	51.7	5.41
P_{out}	7.2 MW	2.0 MW

Due to operation near the synchronous cavity length in the current example ($\delta L_c = \delta L_c^p/3$), the net round trip gains G_{net} in step (4) are quite small. These can be increased by reducing the cavity losses with only a minor effect on the output optical power (recall figure 11.6). If we operate with $\delta_c = \frac{1}{2}\delta_c^{\text{opt}}$ for both harmonics, then $G_{\text{net};1} = 65.3\%$ and $G_{\text{net};3} = 36.3\%$. The corresponding output powers are calculated by the procedure described in the text following (11.18), with $a' \to a_h'$ formally, using the parameters $G' \to G_\gamma'$; $4\pi \to \alpha_\gamma$; $\sqrt{\pi} \to \beta_\gamma$ from section 11.6. Upon converting $|a_h'|^2_{\text{out}}$ to $|a_p|^2_{\text{out}}$ we calculate $P_1^{\text{out}} = 6.4$ MW; $P_3^{\text{out}} = 1.8$ MW.

12.3 Spontaneous emission

Although the FEL wave equation has been derived and discussed in the context of laser amplification, it is enlightening to use it to calculate spontaneous emission into arbitrary harmonics; the general result is given in section 5.3. Start with the wave equation for the fth harmonic, (12.6), where j_F is given in (12.9). We assume that the electron beam is filamentary and calculate the spontaneous power emitted during a short time interval $\Delta\tau$ within the collimated waist of the fundamental resonator mode; thus, we set $q_E = 1$.

Consider a monoenergetic, rectangular electron bunch with a bunch length equal to one fundamental wavelength λ. Let the peak current be I so that the number of electrons, randomly distributed in phase, is

$$N_e = \frac{(I/e)_{\text{MKS}}\,\lambda}{c}. \tag{12.15}$$

The spontaneous field emitted by these electrons into the vacuum mode is obtained by integrating (12.6) over a time duration $\Delta\tau = 1/N_w$; the resulting field increment $\Delta a_h'$ slips ahead of the electrons and into the vacuum after that time. The spontaneous field emitted by the electron bunch is then

$$\Delta a_h' = a_h' = -fj_F\,\Delta\tau\cdot\langle e^{-i\xi'}\rangle_{\xi_0'} = -\frac{fj_F}{N_w}\cdot\left[\sum_{n=1}^{Ne} e^{-i\xi_n'}\right]/N_e, \tag{12.16}$$

and the optical power in the case of incoherent emission is

$$|a_h'|^2 = \frac{f^2 j_F^2}{N_w^2 N_e}, \tag{12.17}$$

where cross terms in the mod-squared of the summation dropped out for random phases ξ_n'. Using (12.13) to convert from $|a_h'|^2$ to $|a_p|^2$, and (12.9) and (12.15) to substitute for j_F and N_e, we calculate the spontaneous optical power from (11.27) to be (in CGS units)

$$P_{\rm op} = \frac{\gamma^4 m^2 c^5}{32\pi^3 e^2 N_w^4 \lambda_w^2 \hat{K}_f^2} \cdot \frac{\pi w_0^2}{2} \cdot \frac{j_F^2}{2N_w^2 N_e} \tag{12.18}$$

$$= \frac{\gamma^4 m^2 c^5}{32\pi^3 e^2 N_w^4 \lambda_w^2 \hat{K}_f^2} \cdot A_{\rm opt} \cdot \frac{1}{2N_w^2} \frac{c}{(I/e)_{\rm MKS}\,\lambda} \cdot \frac{64\pi^4 e^4 N_w^6 \lambda_w^4 \hat{K}_f^4 (I/e)_{\rm MKS}^2}{m^2 c^6 \gamma^6 A_{\rm opt}^2} \tag{12.19}$$

$$= \frac{\pi e^2 \lambda_w^2 \hat{K}_f^2 (I/e)_{\rm MKS}}{\gamma^2 A_{\rm opt} \lambda}. \tag{12.20}$$

We convert from optical power to 'emitted' power by multiplying by $\frac{\lambda}{\lambda_W}$, as explained in section 5.2:

$$P_e^{\rm bunch} = \frac{\pi e^2 \lambda_w \hat{K}_f^2 (I/e)_{\rm MKS}}{\gamma^2 A_{\rm opt}}, \tag{12.21}$$

which gives the incoherent power emitted by the electron bunch during its passage through the waist of the mode. The power emitted by a single electron is obtained by dividing $P_e^{\rm bunch}$ by N_e from (12.15), yielding

$$P_e = \frac{\pi e^2 \lambda_w \hat{K}_f^2 c}{\gamma^2 A_{\rm opt} \lambda} = \frac{2\pi e^2 \lambda_w \hat{K}_f^2 c}{\gamma^2 \pi w_0^2 \lambda}. \tag{12.22}$$

Since we obtained this power starting from the wave equation with filling factor projected onto the lowest order transverse mode, it represents the power emitted by a single electron into that mode. We can therefore obtain the power emitted per unit solid angle by dividing P_e by the solid angle $\Delta\Omega_f$ given in (5.18), where $\Delta\Omega_f$

is defined for the lowest order transverse mode at the harmonic wavelength $\lambda_f = \lambda/f$:

$$\left\langle \frac{\mathrm{d}P_e}{\mathrm{d}\Omega} \right\rangle_f = \frac{2\pi e^2 \lambda_w \hat{K}_f^2 c}{\gamma^2 \pi w_0^2 \lambda} \left(\frac{\pi w_0^2}{2\lambda_f^2} \right) = \frac{\pi e^2 \lambda_w \hat{K}_f^2 c f^2}{\gamma^2 \lambda^3} = \frac{\pi e^2 \lambda_w \hat{K}_f^2 c f^2}{\gamma^2} \frac{8\gamma^6}{\lambda_w^3 \left(1 + \hat{K}^2\right)^3} \tag{12.23}$$

or
$$\left\langle \frac{\mathrm{d}P_e}{\mathrm{d}\Omega} \right\rangle_f = \frac{f^2 e^2 c K^2 k_w^2 \gamma^4}{\pi \left(1 + \hat{K}^2\right)^3} \left[J_{\frac{f-1}{2}}(f\eta) - J_{\frac{f+1}{2}}(f\eta) \right]^2. \tag{12.24}$$

The reader can confirm that this is precisely the power per unit solid angle given in (5.21). This calculation thus demonstrates a high degree of internal consistency throughout the entire classical FEL analysis.

Chapter 13

Helical undulators

13.1 Electron trajectories

Although most contemporary FELs employ plane-polarized undulators, the original FEL used a helical undulator constructed of a bi-filar winding, with a high dc current traveling in opposite directions in each winding. The axial *B*-fields thus canceled along the axis, leaving only a circularly polarized transverse component. The helical electron trajectory and the corresponding circularly polarized optical field are illustrated in figure 13.1.

Notice that the sense of spatial helicity of the optical field is opposite to the helicity of the electron trajectory and transverse undulator field. This may appear

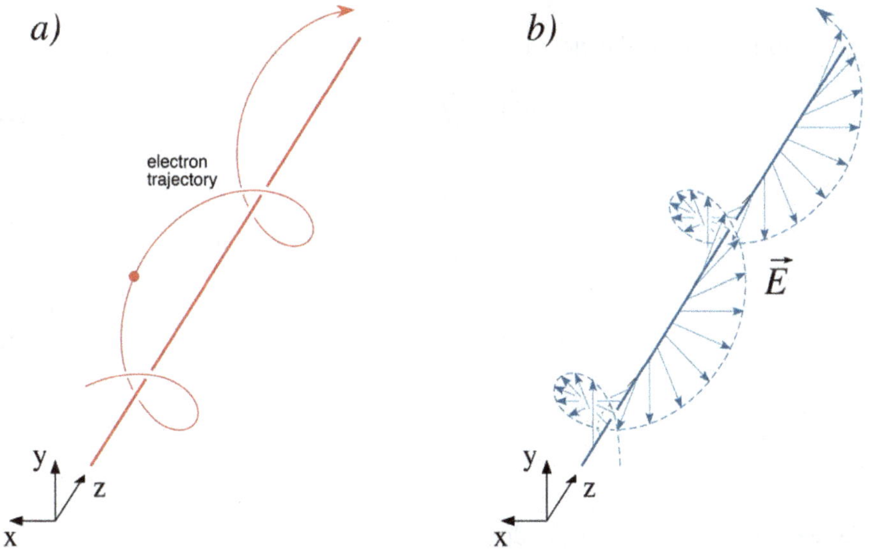

Figure 13.1. (a) Helical electron trajectory and (b) circularly polarized optical field in a helical undulator.

doi:10.1088/978-1-6270-5573-4ch13

counter-intuitive, but the way to check it is to consider the electron motion in the ERF. If the undulator field and electron trajectory have right-threaded helicity in the lab frame, then the electrons will circulate clockwise in the ERF (looking along the direction of motion). This clockwise dipolar motion produces a clockwise electric field at the position of the electron, which requires the spatial sense of the optical helicity to be left-threaded as the optical wave slips ahead of the electrons.

Following the analysis in chapter 4, let us use the spatial components of the Lorentz force equation to derive the electron trajectory in the absence of the optical field. The undulator field corresponding to the helicity shown in figure 13.1(*a*) is

$$\vec{B}_w = \hat{x}\, B_w \cos k_w z + \hat{y}\, B_w \sin k_w z, \tag{13.1}$$

and the $\hat{x}$ and $\hat{y}$ components of the Lorentz equation, (4.1), yield

$$\frac{\mathrm{d}(\gamma\beta_x)}{\mathrm{d}t} = +\frac{e}{mc}\beta_z B_w \sin k_w z; \qquad \frac{\mathrm{d}(\gamma\beta_y)}{\mathrm{d}t} = -\frac{e}{mc}\beta_z B_w \cos k_w z \tag{13.2}$$

$$\frac{\mathrm{d}(\gamma\beta_x)}{\mathrm{d}z} = +\frac{e}{mc^2}B_w \sin k_w z \qquad \frac{\mathrm{d}(\gamma\beta_y)}{\mathrm{d}z} = -\frac{e}{mc^2}B_w \cos k_w z \tag{13.3}$$

$$\beta_x = -\frac{K}{\gamma}\cos k_w z \qquad \beta_y = -\frac{K}{\gamma}\sin k_w z \tag{13.4}$$

where the undulator parameter K has the same form in terms of the peak field B_w as in (4.8),

$$K = \frac{eB_w\lambda_w}{2\pi mc^2}. \tag{13.5}$$

The $\hat{z}$ component of the electron velocity is derived from

$$\frac{1}{\gamma^2} = 1 - \left(\beta_x^2 + \beta_y^2 + \beta_z^2\right) \tag{13.6}$$

$$\beta_z^2 = 1 - \frac{1}{\gamma^2} - \beta_x^2 - \beta_y^2 \tag{13.7}$$

$$\beta_z^2 = 1 - \frac{1}{\gamma^2} - \frac{K^2}{\gamma^2}, \tag{13.8}$$

or $$\beta_z = 1 - \frac{1+K^2}{2\gamma^2}. \tag{13.9}$$

We see that there is no z-dependence either in either β_z or the $\hat{z}$ coordinate of the electron in the lab frame. Therefore, there is no 'figure-of-eight' motion containing higher harmonics of the fundamental frequency as in the plane-polarized undulator, and *no harmonic emission in a helical undulator*. Substitution of β_z from (13.9) into the slippage condition, (1.5), yields a resonant optical wavelength of

$$\lambda = \frac{\lambda_w}{2\gamma^2}\left(1 + K^2\right). \tag{13.10}$$

This differs from the result for the plane-polarized undulator, (4.16), in the appearance of the peak value K in place of the rms value $\hat{K}$.

13.2 FEL coupled equations of motion

The circularly polarized optical field shown in figure 13.1(*b*) is given by

$$\vec{E}_r = \hat{x}\,|E|\cos(kz - \omega t + \phi) - \hat{y}\,|E|\sin(kz - \omega t + \phi). \tag{13.11}$$

Following the analysis in chapter 6, the temporal component of the Lorentz equation is developed as

$$\frac{\mathrm{d}\gamma}{\mathrm{d}t} = -\frac{e}{mc}\vec{\beta}\cdot\vec{E}_r = -\frac{e|E|}{mc}\left[-\frac{K}{\gamma}\cos(k_w z)\cos(kz - \omega t + \phi) + \frac{K}{\gamma}\sin(k_w z)\sin(kz - \omega t + \phi)\right] \tag{13.12}$$

$$= +\frac{eK|E|}{2\gamma mc}\left[\cos(k_w z + kz - \omega t + \phi) + \cos(-k_w z + kz - \omega t + \phi) - \cos(-k_w z + kz - \omega t + \phi) + \cos(k_w z + kz - \omega t + \phi)\right] \tag{13.13}$$

$$= +\frac{eK|E|}{\gamma mc}\cos(k_w z + kz - \omega t + \phi). \tag{13.14}$$

As in (6.7) and (6.8), we define the electron phase ξ and phase velocity ν in the ponderomotive potential,

$$\xi = (k + k_w)z - \omega t \tag{13.15}$$

$$\nu = L_w\left[(k + k_w)\beta_z - k\right]. \tag{13.16}$$

The relationship between $\mathrm{d}\gamma/\mathrm{d}t$ and $\mathrm{d}\nu/\mathrm{d}\tau$ is then the same as in (6.16) and we can rewrite (13.14) as

$$\frac{\mathrm{d}\nu}{\mathrm{d}\tau} = \frac{4\pi e N_w^2 \lambda_w K}{\gamma^2 mc^2}|E|\cos(\xi + \phi) \tag{13.17}$$

$$\frac{\mathrm{d}\nu}{\mathrm{d}\tau} \equiv |a|\cos(\xi + \phi). \tag{13.18}$$

This is the pendulum equation for helical FELs and is identical to the pendulum equation for plane-polarized FELs except for the appearance of the peak values for K and $|E|$ and the absence of the Bessel function factor $[J_n - J_{n+1}]$ in the definition of the dimensionless optical field $a \equiv |a|e^{i\phi}$.

To derive the FEL wave equation following the procedure in chapter 7, we employ an optical vector potential of the form

$$\vec{A} = \hat{x}\,\frac{|E|}{k}\sin(kz - \omega t + \phi) + \hat{y}\,\frac{|E|}{k}\cos(kz - \omega t + \phi), \qquad (13.19)$$

which yields in the SVEA the same field we assumed in (13.11). This vector potential can also be written in complex form as

$$\vec{A} = (\hat{x} + i\hat{y})\frac{E}{2ik}e^{i(kz-\omega t)} + c.c., \qquad (13.20)$$

where $E = |E|e^{i\phi}$ is the complex slowly varying envelope. Imposing the SVEA and the paraxial wave approximation, we obtain the wave equation in the form of (7.12) to be

$$(\hat{x} + i\hat{y})\left[\frac{1}{2ik}\nabla_\perp^2 E + \left(\frac{\partial E}{\partial z} + \frac{1}{c}\frac{\partial E}{\partial t}\right)\right] + (\hat{x} + i\hat{y})[...]^* e^{-2i(kz-\omega t)} = -\frac{4\pi}{c}\vec{J}_\perp\, e^{-i(kz-\omega t)}. \qquad (13.21)$$

With reference to (13.4), note that the transverse current $\vec{J}_\perp$ for the ith electron can be written

$$\vec{J}_\perp \equiv -e\vec{v}_\perp\delta^{(3)}(\vec{r} - \vec{r}_i) = +\frac{ecK}{\gamma}(\hat{x}\cos k_w z + \hat{y}\sin k_w z)\delta^{(3)}(\vec{r} - \vec{r}_i) \qquad (13.22)$$

$$= \frac{ecK}{2\gamma_i}\left[(\hat{x} + i\hat{y})e^{-ik_w z_i} + (\hat{x} - i\hat{y})e^{+ik_w z_i}\right]\delta^{(3)}(\vec{r} - \vec{r}_i). \qquad (13.23)$$

Equation (13.21) then becomes

$$(\hat{x} + i\hat{y})\left[\frac{1}{2ik}\nabla_\perp^2 E + \left(\frac{\partial E}{\partial z} + \frac{1}{c}\frac{\partial E}{\partial t}\right)\right] + (\hat{x} + i\hat{y})[...]^* e^{-2i(kz-\omega t)}$$

$$= -\frac{2\pi eK}{\gamma_i}\left[(\hat{x} + i\hat{y})e^{-ik_w z_i} + (\hat{x} - i\hat{y})e^{+ik_w z_i}\right]\delta^{(3)}(\vec{r} - \vec{r}_i)e^{-i(kz_i-\omega t)} \qquad (13.24)$$

$$= -\frac{2\pi eK}{\gamma_i}\left[(\hat{x} + i\hat{y})e^{-ik_w z_i}e^{-i(kz_i-\omega t)} + (\hat{x} - i\hat{y})e^{+ik_w z_i}e^{-i(kz_i-\omega t)}\right]\delta^{(3)}(\vec{r} - \vec{r}_i). \qquad (13.25)$$

We now perform an average over one magnet period to eliminate the fast oscillating terms (the conjugated term on the lhs and the second term on the rhs) (cf (7.16)) and drive the wave with electrons of density n_e in volume ΔV (cf (7.20)) to obtain

$$\frac{1}{2ik}\nabla_\perp^2 E + \left(\frac{\partial E}{\partial z} + \frac{1}{c}\frac{\partial E}{\partial t}\right) = -2\pi e K n_e \left\langle \frac{e^{-i\xi}}{\gamma} \right\rangle_{\Delta V}. \tag{13.26}$$

For plane waves and small $\Delta\gamma$, this reduces in co-moving coordinates to the dimensionless wave equation (cf (7.29))

$$\frac{da}{d\tau} = -j\langle e^{-i\xi}\rangle_{\xi_0,\nu_0}; \quad j = \frac{8\pi^2 e^2 N_w^3 \lambda_w^2 K^2}{\gamma^3 mc^2} n_e. \tag{13.27}$$

This is the dimensionless wave equation for helical FELs and is again identical to the form obtained for plane-polarized undulators, except for the appearance of the peak value for K and the absence of the Bessel function factor $[J_n - J_{n+1}]$ in the definition of the dimensionless current density j.

13.3 Small-signal gain

Due to the derivation and form of the coupled equations of motion for the helical FEL,

$$\frac{da}{d\tau} = -j\langle e^{-i\xi}\rangle_{\xi_0,\nu_0} \tag{13.28}$$

$$\frac{d\nu}{d\tau} = |a|\cos(\xi + \phi), \tag{13.29}$$

it is clear that all previous results can be applied to the analysis of the helical FEL, both in the small signal regime and at saturation, including any modifications and extensions of the equations to motion, except of course for the absence of harmonic emission. Thus, for example, the small-signal gain of the helical FEL for large currents is given by

$$G' = 0.135 j_F + 0.00486 j_F^2 + 0.0000172 j_F^3, \tag{13.30}$$

where j_F incorporates the filling factor and G' may be further modified by inhomogeneous gain reduction, and so on. However, it is instructive to compare the helical FEL to the plane-polarized FEL more directly to examine the effect of the missing Bessel function factor explicitly.

There are a couple of questions we can ask.

First, what value of K^2 maximizes the gain at a fixed wavelength λ in each type of undulator? Second, if we could choose between a helical or plane-polarized FEL of the same N_w and λ_w, which choice would yield greater gain for operation at common specified values of the wavelength λ and electron energy γ, and how much greater would the gain be? Assume that $G = 0.135j$ to answer these questions.

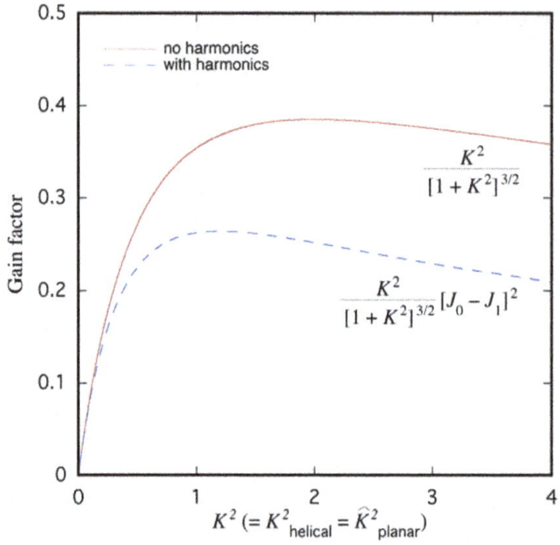

Figure 13.2. Comparison of gain in a helical versus plane-polarized FEL, for common values of λ_w, N_w, λ, γ.

To answer the first question, note that increasing K^2 to increase the gain also requires that we increase γ to maintain operation at the specified wavelength, and these two variations play off against each other in the formula for the gain.

Helical FEL. In light of (13.10), the full K^2 dependence in the gain of the helical FEL is given by

$$G_{\text{helical}} \propto \frac{K^2}{\gamma^3} \propto \frac{K^2}{\left[1+K^2\right]^{\frac{3}{2}}}. \tag{13.31}$$

This is maximized for a value of $K^2 = 2$ (figure 13.2).

Plane-polarized FEL. For the plane-polarized FEL, the complete $\hat{K}^2$ dependence in the gain is given by

$$G_{\text{planar}} \propto \frac{\hat{K}_f^2}{\gamma^3} \propto \frac{\hat{K}^2}{\left[1+\hat{K}^2\right]^{\frac{3}{2}}}\left[J_0\left(\frac{\hat{K}^2}{2\left(1+\hat{K}^2\right)}\right) - J_1\left(\frac{\hat{K}^2}{2\left(1+\hat{K}^2\right)}\right)\right]^2. \tag{13.32}$$

This is maximized for a value of $\hat{K}^2 = 1.2$ (figure 13.2).

If the undulator period and wavelength λ were specified to be common design parameters in each type of undulator (although they wouldn't have to be to have answered our first question), then the peak magnetic field in the plane-polarized undulator would have to be 9.5% higher than the peak helical *B*-field when both were operating at maximum gain. On the other hand, the plane-polarized undulator would be able to operate with an energy γ that was 14% lower.

To answer the second question, note that if λ and γ are both fixed, (13.10) and (4.16) require the undulators to be designed such that

$$K_{\text{helical}}^2 = \hat{K}_{\text{planar}}^2. \tag{13.33}$$

Since

$$G_{\text{helical}} \propto K_{\text{helical}}^2 \qquad \text{and} \qquad G_{\text{planar}} \propto \hat{K}_{f;\text{planar}}^2 \tag{13.34}$$

with the constants of proportionality otherwise being the same, we see that the gain in the plane-polarized FEL is lower than the gain in the helical FEL by a factor of $[J_0 - J_1]^2$.

Chapter 14

Small-signal gain—second derivation

14.1 The equation for weak fields

The derivation of the FEL equation for weak fields in this section is based on the paper by Colson and Blau (1987). We proceed by solving for the evolution of the optical wave directly from the wave equation, after eliminating the electron beam quantities using the pendulum equation. Start with the coupled Maxwell–Lorentz equations, (7.30) and (7.31), and assume formally that $\langle\ldots\rangle_{\xi 0,\nu 0} = \langle\langle\ldots\rangle_{\xi 0}\rangle_{\nu 0}$:

$$\frac{\mathrm{d}^2\xi}{\mathrm{d}\tau^2} = |a|\cos(\xi + \phi) \tag{14.1}$$

$$\frac{\mathrm{d}a}{\mathrm{d}\tau} = -j\langle \mathrm{e}^{-\mathrm{i}\xi}\rangle_{\xi_0,\nu_0}. \tag{14.2}$$

From the pendulum equation, integrate successively to obtain

$$\frac{\mathrm{d}\xi}{\mathrm{d}\tau} = \nu_0 + \int_0^\tau \mathrm{d}q|a(q)|\cos[\xi(q) + \phi(q)] \tag{14.3}$$

$$\xi(\tau) = \xi_0 + \nu_0\tau + \int_0^\tau \mathrm{d}\tau' \int_0^{\tau'} \mathrm{d}q|a(q)|\cos[\xi(q) + \phi(q)] \tag{14.4}$$

$$= \xi_0 + \nu_0\tau + \int_0^\tau \mathrm{d}\tau' \int_0^{\tau'} \mathrm{d}q|a(q)|\cos[\xi_0 + \nu_0 q + \phi(q) + O\{|a|\}] \tag{14.5}$$

$$= \xi_0 + \nu_0\tau + \int_0^\tau \mathrm{d}\tau' \int_0^{\tau'} \mathrm{d}q|a(q)|\cos[\xi_0 + \nu_0 q + \phi(q)] + O\{|a|^2\}, \tag{14.6}$$

doi:10.1088/978-1-6270-5573-4ch14

where we performed a perturbation expansion in the last line. Now substitute this expression for $\xi(\tau)$ into the wave equation:

$$\frac{\mathrm{d}a}{\mathrm{d}\tau} = -j\langle \cos \xi(\tau) - \mathrm{i} \sin \xi(\tau)\rangle_{\xi_0,\nu_0} \tag{14.7}$$

$$= -j\Bigg\langle \cos\left[\xi_0 + \nu_0\tau + \int_0^\tau \mathrm{d}\tau' \int_0^{\tau'} \mathrm{d}q|a(q)| \cos[\xi_0 + \nu_0 q + \phi(q)]\right]$$

$$- \mathrm{i} \sin\left[\xi_0 + \nu_0\tau + \int_0^\tau \mathrm{d}\tau' \int_0^{\tau'} \mathrm{d}q|a(q)| \cos[\xi_0 + \nu_0 q + \phi(q)]\right]\Bigg\rangle_{\xi_0,\nu_0} \tag{14.8}$$

$$= -j\Bigg\langle \cos\left[\xi_0 + \nu_0\tau\right] \cdot (1 + O\{|a|^2\}) - \sin\left[\xi_0 + \nu_0\tau\right]$$

$$\cdot \int_0^\tau \mathrm{d}\tau' \int_0^{\tau'} \mathrm{d}q|a(q)| \cos[\xi_0 + \nu_0 q + \phi(q)]$$

$$- \mathrm{i} \sin\left[\xi_0 + \nu_0\tau\right] \cdot (1 + O\{|a|^2\}) - \mathrm{i} \cos\left[\xi_0 + \nu_0\tau\right]$$

$$\cdot \int_0^\tau \mathrm{d}\tau' \int_0^{\tau'} \mathrm{d}q|a(q)| \cos[\xi_0 + \nu_0 q + \phi(q)]\Bigg\rangle_{\xi_0,\nu_0} \tag{14.9}$$

$$= ij\Bigg\langle \cos\left[\xi_0 + \nu_0\tau\right] \cdot \int_0^\tau \mathrm{d}\tau' \int_0^{\tau'} \mathrm{d}q|a(q)| \cos[\xi_0 + \nu_0 q + \phi(q)]$$

$$- \mathrm{i} \sin\left[\xi_0 + \nu_0\tau\right] \cdot \int_0^\tau \mathrm{d}\tau' \int_0^{\tau'} \mathrm{d}q|a(q)| \cos[\xi_0 + \nu_0 q + \phi(q)]\Bigg\rangle_{\xi_0,\nu_0}, \tag{14.10}$$

where the average over $\langle ... \rangle_{\xi_0}$ in the first and third terms in (14.9) dropped out. Move the $\cos[\xi_0 + \nu_0\tau]$ and $\sin[\xi_0 + \nu_0\tau]$ factors into the respective integrals and expand the trig products using $\cos A \cos B = ...$, $\sin A \cos B = ...$ The average over $\langle ... \rangle_{\xi_0}$ can be evaluated exactly and we are left with $\langle ... \rangle_{\nu_0}$:

$$\frac{\mathrm{d}a}{\mathrm{d}\tau} = \mathrm{i}j \int_0^\tau \mathrm{d}\tau' \int_0^{\tau'} \mathrm{d}q|a(q)|$$

$$\left\langle \frac{1}{2} \cos\left[\nu_0(\tau - q) - \phi(q)\right] - \frac{\mathrm{i}}{2} \sin\left[\nu_0(\tau - q) - \phi(q)\right]\right\rangle_{\nu_0} \tag{14.11}$$

$$= \frac{\mathrm{i}j}{2} \int_0^\tau \mathrm{d}\tau' \int_0^{\tau'} \mathrm{d}q|a(q)|\left\langle \mathrm{e}^{-\mathrm{i}[\nu_0(\tau-q)-\phi(q)]}\right\rangle_{\nu_0} \tag{14.12}$$

$$= \frac{\mathrm{i}j}{2} \int_0^\tau \mathrm{d}\tau' \int_0^{\tau'} \mathrm{d}q \cdot a(q) \cdot \left\langle \mathrm{e}^{-\mathrm{i}\nu_0(\tau-q)}\right\rangle_{\nu_0}. \tag{14.13}$$

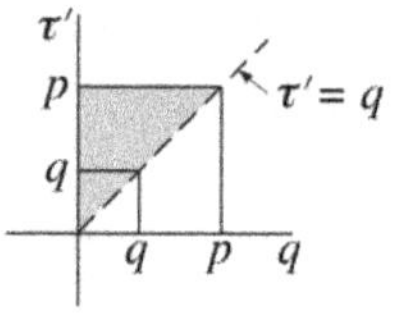

Figure 14.1. Region of integration in (14.14).

Integrate one more time,

$$a(\tau) = a_0 + \frac{\mathrm{i}j}{2}\int_0^{\tau} \mathrm{d}p \int_0^{p} \mathrm{d}\tau' \int_0^{\tau'} \mathrm{d}q \cdot a(q) \cdot \left\langle \mathrm{e}^{-\mathrm{i}\nu_0(p-q)} \right\rangle_{\nu_0}, \tag{14.14}$$

and interchange the order of integration over $\{\mathrm{d}\tau', \mathrm{d}q\}$ to obtain (figure 14.1)

$$a(\tau) = a_0 + \frac{\mathrm{i}j}{2}\int_0^{\tau} \mathrm{d}p \int_0^{p} \mathrm{d}q \left[\int_q^{p} \mathrm{d}\tau'\right] a(q) \cdot \left\langle \mathrm{e}^{-\mathrm{i}\nu_0(p-q)} \right\rangle_{\nu_0} \tag{14.15}$$

$$a(\tau) = a_0 + \frac{\mathrm{i}j}{2}\int_0^{\tau} \mathrm{d}p \int_0^{p} \mathrm{d}q \cdot (p-q) \cdot a(q) \cdot \left\langle \mathrm{e}^{-\mathrm{i}\nu_0(p-q)} \right\rangle_{\nu_0}. \tag{14.16}$$

This is the solution for the time-dependent optical field in the small signal regime. The optical field at time τ depends on the optical field at earlier times q, because of optical slippage and the interaction between the optical wave and the electrons: the optical field at time q disturbs the electron distribution, and then slips over those electrons, which in turn feed back on the optical field at later times τ.

If the energy spread in the electron beam is represented by a Gaussian distribution over ν_0, centered on ν_p as in (10.22),

$$\rho(\nu_0) = \frac{1}{\sqrt{2\pi\sigma^2}} \exp\left[-\frac{(\nu_0 - \nu_p)^2}{2\sigma^2}\right], \tag{14.17}$$

then the remaining average over ν_0 in (14.16) is given by

$$\left\langle \mathrm{e}^{-\mathrm{i}\nu_0(p-q)} \right\rangle_{\nu_0} = \int_{\nu_0} \mathrm{e}^{-\mathrm{i}\nu_0(p-q)} \rho(\nu_0)\mathrm{d}\nu_0 \tag{14.18}$$

$$= \frac{1}{\sqrt{2\pi\sigma^2}} \int_{-\infty}^{\infty} \mathrm{e}^{-(\nu_0-\nu_p)^2/2\sigma^2} \mathrm{e}^{-\mathrm{i}\nu_0(p-q)}\, \mathrm{d}\nu_0 \tag{14.19}$$

$$= \frac{1}{\sqrt{2\pi\sigma^2}} \mathrm{e}^{-\mathrm{i}\nu_p(p-q)} \int_{-\infty}^{\infty} \mathrm{e}^{-\nu^2/2\sigma^2} \mathrm{e}^{-\mathrm{i}\nu(p-q)}\, \mathrm{d}\nu \tag{14.20}$$

$$= \mathrm{e}^{-\mathrm{i}\nu_p(p-q)} \mathrm{e}^{-\sigma^2(p-q)^2/2}, \tag{14.21}$$

where we made the variable substitution $\nu = \nu_0 - \nu_p$ in the third line and evaluated the resulting integral using that most useful of integration formulas (Siegman 1986)

$$\int_{-\infty}^{\infty} \mathrm{e}^{-ay^2}\mathrm{e}^{-2by}\,\mathrm{d}y = \sqrt{\frac{\pi}{a}}\,\mathrm{e}^{b^2/a}; \qquad \mathcal{Re}\{a\} > 0. \tag{14.22}$$

14.2 FEL gain and dispersion

To calculate the small-signal gain, we explicitly impose the small gain condition $a(q) \rightarrow a_0$ and factor the field from the integral. Thus, upon integrating to $\tau = 1$ we obtain for a monoenergetic electron beam

$$\frac{a(1) - a_0}{a_0} = \frac{\mathrm{i}j}{2}\int_0^1 \mathrm{d}p \int_0^p \mathrm{d}q \cdot (p - q) \cdot \mathrm{e}^{-\mathrm{i}\nu_0(p-q)}. \tag{14.23}$$

The successive integrals can be evaluated directly and we calculate a complex gain of

$$g \equiv \frac{a(1) - a_0}{a_0} = j\left[\frac{2 - 2\cos\nu_0 - \nu_0\sin\nu_0}{2\nu_0^3} + \mathrm{i}\,\frac{2\sin\nu_0 - \nu_0 - \nu_0\cos\nu_0}{2\nu_0^3}\right]. \tag{14.24}$$

Not only is there a real part responsible for gain, but also an imaginary part responsible for refractive effects. These curves are plotted in figure 14.2. It can also

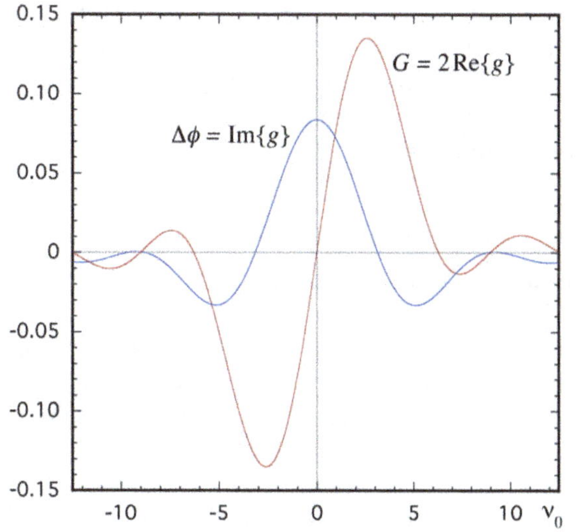

Figure 14.2. The complex small-signal gain function. G and $\Delta\phi$ are the power gain and phase shift of the optical wave on a single pass ($\tau = 0 \rightarrow 1$).

be shown that the real and imaginary parts satisfy causality and are related by a Hilbert transform. The power gain G is related to the amplitude gain g by

$$G \equiv \frac{|a(1)|^2 - |a_0|^2}{|a_0|^2} = \left|\frac{a(1)}{a_0}\right|^2 - 1$$

$$= |g + 1|^2 - 1 = g + g^* + |g|^2; \quad \text{or} \quad G = 2\,\mathcal{Re}\{g\}, \tag{14.25}$$

where we neglect the small term $|g|^2$ in the expressly small gain regime. Therefore, we have

$$G = j \cdot \left[\frac{2 - 2\cos\nu_0 - \nu_0 \sin\nu_0}{\nu_0^3}\right], \tag{14.26}$$

which is the same as we obtained in chapter 9, (9.21). The phase shift of the optical wave on a single pass is

$$\Delta\phi \equiv \arg\left\{\frac{a(1)}{a_0}\right\} = \arg\{1 + g\}$$

$$= \arg\{1 + \mathcal{Re}\{g\} + \mathrm{i}\,\mathrm{Im}\{g\}\}; \quad \text{or} \quad \Delta\phi = \mathrm{Im}\{g\}, \tag{14.27}$$

where we again employ the small gain condition $|g| \ll 1$. The maximum gain G occurs at $\nu_0 = 2.606$, at which point the associated phase shift is $\Delta\phi = +0.0185j$. Does this phase shift advance or retard the wavefront? Well, our choice of phase for the optical field, on which this text is based, is $\exp \mathrm{i}(kz - \omega t + \Delta\phi)$. If we freeze time, we see that the optical phase increases with increasing z, which moves in a direction ahead of the wavefronts. We conclude that if z is held fixed, a positive value for $\Delta\phi$ moves the wavefront *back*.

So, the FEL interaction retards the co-propagating wavefronts. Since this effect is proportional to the current density j, the refractive effect of an electron beam localized on the axis is to focus the optical beam. This is called *refractive guiding* and has important implications for counteracting diffraction and increasing the interaction length in long undulators.

14.3 A digression on numerical simulations

In FEL simulations in which slippage is explicitly included, it is often expeditious to choose the optical bin size in the numerical window to equal the resonant optical wavelength λ determined by the incident energy γ in (4.16). Slippage is then simply incorporated, with good numerical accuracy, by setting the total number of integration time steps equal to the number of undulator periods. With this prescription, the optical field is shifted by a single bin after each time step to account for slippage between the electrons and optical wave.

The optical frequency is then determined by the microscopic phase of the optical field, which varies from bin to bin and is affected both by the FEL interaction and by slippage. At a given position $\tilde{z}$ in the optical window and at time τ in the interaction, the *deviation* from the resonant frequency ω_r is $\omega'(\tilde{z}, \tau) = c\, \partial\phi(\tilde{z}, \tau)/\partial\tilde{z}$, or $\omega_n' = \nu_r \Delta_n \phi$ for the nth bin, where $\nu_r = c/\lambda$; more directly, we calculate the discrete Fourier transform of the optical field to determine the spectral width and offset from the resonant frequency ω_r.

It may turn out that the frequency offset ω' due to frequency pulling is too large a fraction of the numerical spectral window to maintain good numerical accuracy in the discrete Fourier transform. In this case, the frequency offset is adjusted by specifying a *separate energy offset* $\delta\gamma$ to impose a spectral shift $\delta\omega = -\omega'$ that re-zeros the spectral offset. For example, if $\omega' < 0$ due to frequency pulling, then (11.4) requires a positive energy shift of $\delta\gamma/\gamma \equiv +\frac{1}{2}\delta\omega/\omega_r = -\frac{1}{2}\omega'/\omega_r$ to shift ω' back in the positive direction. (Note that the phase velocity v in (11.4) is fixed by the FEL interaction, so $\delta\nu = 0$ for this adjustment.)

The actual incident energy in the simulation is then $\gamma_i = \gamma + \delta\gamma$ and the actual optical frequency in the simulation is $\omega = \omega_r + \omega'_{\text{resid}}$, where ω'_{resid} is any residual 'uncompensated' frequency offset. In all cases, the bin width remains fixed at the resonant wavelength λ determined by the original energy γ in (4.16).

References

Colson W B and Blau J 1987 Free electron laser theory in weak optical fields *Nucl. Instrum. Methods* A **259** 198–202

Siegman A E 1986 *Lasers* (Mill Valley, CA: University Science Books)

Chapter 15

Short-pulse propagation

15.1 General description

Apart from a summary of results of the supermode theory in section 10.4, all of our analyses thus far have described only CW beam interactions and no short-pulse effects. For example, all dynamic variables so far have been functions only of τ, the 'distance' along the undulator, e.g. $a(\tau)$; $\phi(\tau)$; $\xi(\tau;\, \xi_0,\, \nu_0)$; $\nu(\tau;\, \xi_0,\, \nu_0)$, and describe the FEL interaction without explicit regard to slippage effects or microtemporal position within the beams.

This is clearly not sufficient for the description of short electron and optical pulses: end-effects are obviously important, since the optical pulses can slip over the electron bunches by a substantial fraction of their length. (In the Mark III FEL at 3 μm, the slippage is ~0.5 ps, while the electron bunches are only ~1 to 2 ps long.) Moreover, the modeling of electron shot noise in which individual samples of charge must be tracked, or the simulation of microtemporal effects such as frequency pulling and the sideband instability (see section 11.2), requires optical slippage to be included explicitly.

Historically, an extensive short-pulse theory known as the *supermode theory* was developed by Dattoli and Renieri (1981) starting from the fact that, in an optical resonator, not a single CW optical mode but rather many such modes (the axial modes of the cavity) interact with the electron bunches. These cavity modes become mode locked by the short electron bunches and, under the mode locking effect, evolve towards an 'equilibrium' configuration that retains the same spectral shape from pass to pass, except for an overall complex constant related to laser gain. This mode evolution is illustrated conceptually in figure 15.1.

Mathematically, the interaction and evolution of the axial modes are described by a *coupled-mode matrix* $\mathcal{M}$ whose dimension equals the number of basis-set modes in the problem—these are simply the individual axial modes and there are typically thousands of them—and whose matrix elements quantify the coupling among the modes governed by the FEL interaction and the mode locking effect of the electron bunches.

doi:10.1088/978-1-6270-5573-4ch15

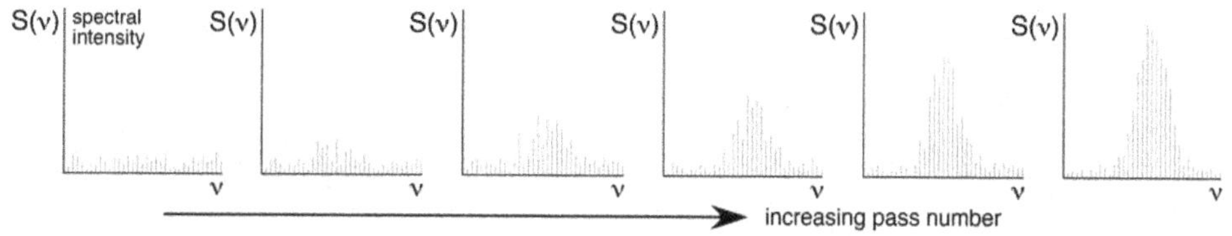

Figure 15.1. Conceptual illustration of the spectral development of an FEL supermode.

The eigenvectors of the coupling matrix are referred to as *supermodes*, each supermode representing a unique configuration of CW axial modes that remains self-similar after one round trip in the cavity. In general, there are as many supermodes (i.e. eigenvectors) as there are dimensions of the coupling matrix, but under successive 'applications' or multiplications of the coupling matrix on repeated round trips in the cavity, the single supermode with the largest eigenvalue (i.e. gain) ultimately dominates the spectrum.

The dominant supermode in the frequency domain is simply the Fourier transform of the stable circulating optical pulse in the time domain, and the corresponding $|\text{eigenvalue}|^2$ yields the net round-trip laser gain. The substructure of discrete modes in the optical spectrum reflects the fact that the stable pulse is repeated every round trip in the cavity, so a comb of modes (the axial mode structure of the cavity) in the frequency domain yields a comb of pulses (i.e. an optical pulse train) in the time domain.

The supermode theory was presented in a series of comprehensive papers in the late 1970s and early 1980s, and was instrumental in elucidating the short-pulse behavior of RF-linac FELs. The analytic solutions can be conveyed in relatively simple formulas of wide applicability (section 10.4). The purpose of the final section of this text is to demonstrate that the supermode theory is analytically equivalent to the FEL description based on the coupled Maxwell–Lorentz equations and can be derived from those equations directly.

15.2 The coupled Maxwell–Lorentz equations

First, we need to generalize the coupled equations of motion to separately account for microtemporal position and slippage within each of the electron and optical pulses.

(i) The optical quantities $\{a, \phi\}$ will be functions of the two independent variables $\tilde{z} \equiv z - ct$ and τ (see figure 15.2(*a*)), where $\tilde{z}$ refers to the position within the optical pulse in a co-moving frame traveling along with the optical pulse at speed c, and τ is the dimensionless time along the undulator, during the interaction.

(ii) The electron coordinates $\{\xi, \nu\}$ will be functions of the two independent variables $\tilde{z}' \equiv z - \bar{v}_z t$ and τ (see figure 15.2(*b*)), where $\tilde{z}'$ refers to the position within the electron bunch in a co-moving frame traveling along with the electron bunch ERF at speed $\bar{v}_z$, and τ is the dimensionless time along the undulator, during the interaction.

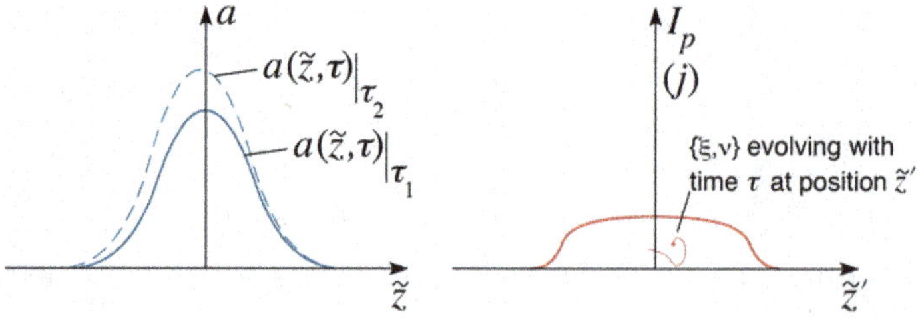

Figure 15.2. (*a*) The optical quantities $\{a, \phi\}$ and (*b*) the electron coordinates $\{\xi, \nu\}$. See text for details.

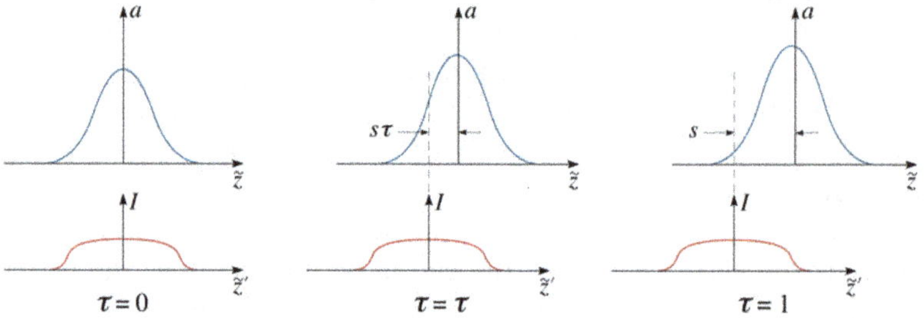

Figure 15.3. Coordinate relations between the electron bunch and optical pulse.

Now if the two reference frames are coincident at the beginning of the undulator ($\tau = 0$), we know that they will be displaced from one another by the slippage distance

$$s \equiv N_w \lambda \tag{15.1}$$

at the end of the undulator ($\tau = 1$), with the optical frame leading the electron frame. At intermediate times τ the displacement will be $s\tau$ (see figure 15.3).

Consider the FEL interaction at time τ between the optical wave at some position within the pulse and the electrons within the same physical volume at the same physical position. By considering the slippage explicitly, we see that at arbitrary τ, the optical field at coordinate $\tilde{z}$ is coincident with the electrons at coordinate $\tilde{z}' = \tilde{z} + s\tau$, and the electrons at coordinate $\tilde{z}'$ are coincident with the optical field at coordinate $\tilde{z} = \tilde{z}' - s\tau$.

To generalize the coupled equations of motion, (14.1) and (14.2), we simply insert these coordinate relations into the respective quantities, yielding for $f = 1$

$$\frac{\partial^2 \xi(\tilde{z}', \tau)}{\partial \tau^2} = |a(\tilde{z}' - s\tau, \tau)| \cos\left[\xi(\tilde{z}', \tau) + \phi(\tilde{z}' - s\tau, \tau)\right] \tag{15.2}$$

$$\frac{\partial a(\tilde{z}, \tau)}{\partial \tau} = -j(\tilde{z} + s\tau)\left\langle \mathrm{e}^{-\mathrm{i}\xi(\tilde{z}+s\tau,\tau)} \right\rangle_{\xi_0, \nu_0}. \tag{15.3}$$

These are the generalized, coupled equations of motion. They are valid for small signals and large signals, for small gain and high gain, for arbitrary electron pulse shapes $j(\tilde{z}')$ and for any magnitude of optical slippage s. They are the starting point for contemporary pulse propagation simulations, *and they contain the entire FEL supermode theory.*

But!—and this is important to appreciate—they contain no new physics beyond what we have already developed. They are the same pendulum and wave equations, and all that the extra coordinate dependencies do is let us keep track of which parts of the optical pulse are interacting with which parts of the electron bunch, and vice versa. Indeed, it is clear from the way we originally derived the equations that their microscopic validity was fundamental.

15.3 Optical pulse evolution

The mathematical notation in (15.2) and (15.3) is rigorous, especially regarding the interpretation of the partial derivatives with respect to the independent variables $\{\tilde{z}', \tau\}$ and $\{\tilde{z}, \tau\}$. We essentially wish to repeat the calculation in section 14.1 to obtain the weak-field solution for $a(\tilde{z}, \tau)$. First, integrate the pendulum equation (15.2) to first order in $|a|$:

$$\frac{\partial\xi(\tilde{z}', \tau)}{\partial\tau} = \nu_0(\tilde{z}') + \int_0^\tau \mathrm{d}q\ |a(\tilde{z}' - sq, q)| \cos\left[\xi(\tilde{z}', q) + \phi(\tilde{z}' - sq, q)\right] \quad (15.4)$$

$$\xi(\tilde{z}', \tau) = \xi_0 + \nu_0(\tilde{z}')\tau + \int_0^\tau \mathrm{d}\tau' \int_0^{\tau'} \mathrm{d}q |a(\tilde{z}' - sq, q)| \cos\left[\xi_0 + \nu_0(\tilde{z}')q + \phi(\tilde{z}' - sq, q)\right] + O\left\{|a|^2\right\}. \quad (15.5)$$

The dependence of ν_0 on $\tilde{z}'$ allows for the possibility that the initial phase velocities may vary along the electron bunch at the beginning of the undulator, as in the chirped-pulse FEL. In the usual case, however, ν_0 is uniform, so for now let us consider a monoenergetic electron beam with $\nu_0(\tilde{z}') \equiv \nu_0$. Before we substitute $\xi(\tilde{z}', \tau)$ into the wave equation, we must evaluate it for $\tilde{z}' = \tilde{z} + s\tau$:

$$\xi(\tilde{z} + s\tau, \tau) = \xi_0 + \nu_0\tau + \int_0^\tau \mathrm{d}\tau' \int_0^{\tau'} \mathrm{d}q \left| a\left(\tilde{z} + s(\tau - q), q\right) \right| \cos\left[\xi_0 + \nu_0 q + \phi\left(\tilde{z} + s(\tau - q), q\right)\right] + O\left\{|a|^2\right\}. \quad (15.6)$$

Then

$$\begin{aligned} \frac{\partial a(\tilde{z}, \tau)}{\partial\tau} &= -j(\tilde{z} + s\tau)\left\langle\cos\xi(\tilde{z} + s\tau, \tau) - \mathrm{i}\sin\xi(\tilde{z} + s\tau, \tau)\right\rangle_{\xi_0} \\ &= \ldots \text{ same steps as section 14.1, keeping track of all variable dependencies} \ldots \end{aligned} \quad (15.7)$$

$$= \frac{\mathrm{i}j(\tilde{z} + s\tau)}{2} \int_0^\tau \mathrm{d}\tau' \int_0^{\tau'} \mathrm{d}q \cdot a\left(\tilde{z} + s(\tau - q), q\right) \cdot \mathrm{e}^{-\mathrm{i}\nu_0(\tau - q)}. \quad (15.8)$$

A final integration, with an interchange in the order of integration over $\{d\tau', dq\}$, yields

$$a(\tilde{z},\tau) = a_0(\tilde{z}) + \frac{i}{2}\int_0^{\tau} dp\, j(\tilde{z}+sp) \int_0^{p} dq \cdot (p-q) \cdot a\big(\tilde{z}+s(p-q), q\big) e^{-i\nu_0(p-q)}. \tag{15.9}$$

This is the optical pulse evolution equation for weak fields, valid for large and small gains, for arbitrary pulse shapes $j(\tilde{z}+sp)$, and for any magnitude of slippage s. It differs from the manifestly small-gain supermode result of Dattoli *et al* (1993) only because it is more general.

The small-gain supermode result (Dattoli *et al* 1993), written in original notation, is:

$$\Delta E(z) = \frac{iK}{2\Delta^3}\int_0^{\Delta} d\eta\, \eta\, e^{-i\nu_0\eta/\Delta} E(z+\eta) \int_{z+\eta}^{z+\Delta} dz' f(z'), \tag{15.10}$$

where $f(z)$ is the longitudinal electron distribution and K is a normalization constant. To obtain this result from (15.9), we impose the small-gain condition by eliminating the explicit time dependence in $a(\tilde{z},\tau)$:

$$a(\tilde{z},\tau) \rightarrow a(\tilde{z}) \tag{15.11}$$

$$a\big(\tilde{z}+s(p-q), q\big) \rightarrow a\big(\tilde{z}+s(p-q)\big). \tag{15.12}$$

Then, replacing $\tilde{z} \rightarrow z$ and integrating to $\tau = 1$ in (15.9), we obtain

$$\Delta a(z) \equiv a_1(z) - a_0(z) = \frac{i}{2}\int_0^{1} dp\, j(z+sp)\int_0^{p} dq \cdot (p-q) \cdot a(z+s(p-q)) e^{-i\nu_0(p-q)} \tag{15.13}$$

$$y \equiv p-q \rightarrow \quad = \frac{i}{2}\int_0^{1} dp\, j(z+sp)\int_0^{p} dy\, y\, a(z+sy) e^{-i\nu_0 y} \tag{15.14}$$

$$\text{interchange} \iint \rightarrow \quad = \frac{i}{2}\int_0^{1} dy\, y\, a(z+sy) e^{-i\nu_0 y}\int_y^{1} dp\, j(z+sp) \tag{15.15}$$

$$sp \equiv q \rightarrow \quad = \frac{i}{2s}\int_0^{1} dy\, y\, a(z+sy) e^{-i\nu_0 y}\int_{sy}^{s} dq\, j(z+q) \tag{15.16}$$

$$sy \equiv \eta \rightarrow \quad = \frac{i}{2s^3}\int_0^{s} d\eta\, \eta\, a(z+\eta) e^{-i\nu_0\eta/s}\int_{\eta}^{s} dq\, j(z+q) \tag{15.17}$$

$$z' \equiv z+q \rightarrow \quad = \frac{i}{2s^3}\int_0^{s} d\eta\, \eta\, a(z+\eta) e^{-i\nu_0\eta/s}\int_{z+\eta}^{z+s} dz'\, j(z'). \tag{15.18}$$

Finally, with a change in notation from $a \to E$ (optical field), $s \to \Delta$ (slippage distance) and $j(z') \to Kf(z')$ (electron distribution), we recover the supermode result of (15.10):

$$\Delta E(z) = \frac{\mathrm{i}K}{2\Delta^3}\int_0^{\Delta} \mathrm{d}\eta\; \eta\; E(z+\eta)\mathrm{e}^{-\mathrm{i}\nu_0\eta/\Delta}\int_{z+\eta}^{z+\Delta} \mathrm{d}z'\, f(z'). \qquad \text{QED} \quad (15.19)$$

15.4 Cavity detuning and refractive effects

The conciseness of (15.19) conceals a wealth of physical implications. One of the results obtained by solving it in the presence of cavity loss and detuning terms with $\Delta \ll \sigma_z$ is that the cavity length yielding the largest gain is shorter than the synchronous length by a displacement of (10.35)

$$\delta L_c^p = -0.0181\, jN_w\lambda. \qquad (15.20)$$

This result can be obtained from fundamental principles of mode-locked laser theory based on our previous CW analysis of the small-signal FEL. Consider a mode locked laser with any form of loss modulation, phase modulation, or gain modulation; RF-linac FELs in particular employ the latter. As explained by Siegman (1986; chapter 27), the cavity modulation induces sidebands on each of the CW axial modes of the cavity, which then injection lock the neighboring axial modes. For optimum mode locking, i.e. maximum gain, the sideband separation—equal to the modulation frequency—must equal the passive (unmodulated) axial mode spacing, so that the sidebands excite the greatest degree of resonance among all of the modes.

From Siegman (1986; chapter 11) the condition that the frequency ω correspond to an axial mode of the cavity in the presence of loss, gain, and any dispersion is that the denominator of the frequency response function

$$\left|\frac{E_{\text{circ}}}{E_{\text{inc}}}\right| = \left|\frac{t}{1 - r_1 r_2\, \mathrm{e}^{-\delta/2}\mathrm{e}^{g}\mathrm{e}^{\mathrm{i}\left(\frac{\omega}{c}\right)2L_c}}\right| \qquad (15.21)$$

correspond to an absolute minimum, where $\mathrm{e}^{-\delta/2}$ is the amplitude reduction due to cavity loss, g is the complex amplitude gain, $r_{1,2}$ are mirror reflectivities and L_c is the cavity length. The resonant frequency of the qth axial mode will thus be obtained if the second term in the denominator is a real maximum. If the mirror reflectivities and cavity loss in the FEL are real, this condition requires

$$\mathrm{Im}\{g(\omega_q)\} + \frac{\omega_q}{c}2L_c = 2\pi q. \qquad (15.22)$$

The spacing between adjacent axial modes $(q, q+1)$ is then

$$\Delta\omega_{\text{ax}} = \omega_{q+1} - \omega_q = \frac{2\pi c}{2L_c + c\dfrac{\mathrm{d}}{\mathrm{d}\omega}\mathrm{Im}\{g\}}. \qquad (15.23)$$

Now, from the discussion of mode locking, the condition for optimum mode locking is that

$$\Delta\omega_{ax} = \omega_{RF} = \text{modulation frequency, fixed by RF-linac.} \tag{15.24}$$

If there were no gain dispersion, this would simply be the condition that the round trip time in the resonator equals the arrival time of the incoming electron bunches (the FEL synchronism condition). But the presence of the second term involving $\frac{d}{d\omega}\mathrm{Im}\{g\}$, related to *group delay*, modifies the cavity length L_c required to maintain synchronism (or resonance, or optimum mode locking). In particular, if the linac frequency ω_{RF} remains fixed, we must impose a mirror displacement of

$$\delta L_c^p = -\frac{1}{2}\,c\,\frac{d}{d\omega}\mathrm{Im}\{g\} \tag{15.25}$$

to compensate that term. Now Im{g} was calculated in section 14.2. Differentiation of this function at the peak of the gain curve $\nu_0 = 2.606$ gives

$$\frac{d}{d\omega}\mathrm{Im}\{g\} = -\frac{N_w\lambda}{c}\frac{d}{d\nu}\Big|_{\nu_0}\mathrm{Im}\{g\} = +j\,\frac{N_w\lambda}{c}\,[0.0363], \tag{15.26}$$

and substitution into (15.25) yields

$$\delta L_c^p = -0.0182\,jN_w\lambda, \tag{15.27}$$

which recovers the result from the supermode theory. This analysis reveals the following:

1) δL_c is essentially independent of both the shape and duration of the electron bunches;
2) the analysis delivers an important bit of physics concerning RF-linac FELs;
3) the result derives from a fundamental property of mode locked lasers as explained by Siegman (1986).

15.5 Mode locked FEL theory

The analytic results of the supermode theory in the frequency domain, specifically the matrix elements of the coupled-mode matrix $\mathcal{M}$, can be derived by applying conventional mode locked laser theory to the coupled FEL equations of motion in the time domain, (15.2) and (15.3), as originally demonstrated by the author (Szarmes 1992, 1993). We first present an overview of mode locking.

The spectral energy distribution of the circulating optical field inside a laser cavity consists, in general, of a superposition of longitudinal cavity modes. For a slowly varying optical field this superposition is written

$$E(t)e^{-i\omega t} = |E(t)|e^{i\phi(t)}e^{-i\omega t} = \sum_n E_n(t)e^{-i\omega_n t} = \sum_n E_n(t)e^{-i2\pi f_n t}, \tag{15.28}$$

in which $E(t)$ is a complex, slowly varying amplitude that encodes the shape and evolution of the intracavity optical field in the time domain (perhaps in the form of a short optical pulse), and the complex amplitudes $E_n(t)$ represent discrete spectral components that have a slowly varying time dependence only over many passes in the cavity. These spectral components $E_n(t)$ essentially describe the long-term spectral evolution of the laser field.

In a free-running laser oscillator, the longitudinal modes coincide with the axial normal modes of the cavity and are separated in frequency by the passive axial mode spacing, which includes frequency pulling effects due to intracavity dispersion. However, in the presence of active intracavity loss-, phase-, or gain-modulation, the longitudinal modes will be separated by the driving frequency of the forced modulation. The laser will reach sustained oscillation only if the modulation frequency is sufficiently close to the passive axial normal mode spacing.

In terms of a time perturbative description of the laser field, the active laser modulation produces sideband components on each of these essentially CW normal modes and these sidebands then injection lock those neighboring normal modes with which they are in resonance. Eventually, the normal modes evolve into a single *hypermode* of coupled sideband components whose spectral center-frequency coincides with the axial normal mode exhibiting the largest gain. This injection locking process continues until all of the coupled modes in the hypermode are phase locked to one or more of their neighbors, with all of the modes separated by the sideband (i.e. modulation) frequency.

The frequency domain analysis of mode locking that describes this injection locking process is applied by Siegman (1986) to the active loss and phase modulation of a conventional laser oscillator. Our derivation of the FEL coupled-mode matrix $\mathcal{M}$ for gain modulation by a pulsed electron beam follows the same analysis.

We start with the slowly varying evolution equation for the nth mode of a laser oscillator, derived by Siegman (1986) and written for the case of intracavity modulation (with the sign of ω determined by our choice $e^{-i\omega t}$) as

$$\frac{dE_n}{dt} + \left[\frac{\tilde{\gamma}_n}{2} - i(\omega_n - \omega_c)\right]E_n = \frac{\Delta E_n^{\text{mod}}}{T_c} \tag{15.29}$$

or as

$$\Delta E_n + \left[\frac{\gamma_n}{2} - iT_c(\omega_n - \omega_c)\right]E_n = \Delta E_n^{\text{mod}}, \tag{15.30}$$

where $E_n(t)$ is the complex amplitude of the nth laser mode, T_c is the cavity round trip time, $\tilde{\gamma}_n$ is the rate of fractional energy loss, γ_n is the fractional energy loss in one round trip and ΔE_n^{mod} is the change in amplitude ΔE_n induced on E_n by the modulation after one round trip. The driving frequency ω_n must lie sufficiently close to one of the axial normal mode frequencies ω_c in order for the injection locking to succeed, which enforces $(\phi_n - \phi_c) \equiv T_c(\omega_n - \omega_c) \ll 2\pi$.

The phasor interpretation of (15.30) is straightforward: the only contributions to the mode E_n arise from the small phase shift $e^{i(\phi_n - \phi_c)}$, the cavity loss $\frac{\gamma_n}{2}E_n$ and the modulation ΔE_n^{mod}. These contributions are illustrated in figure 15.4.

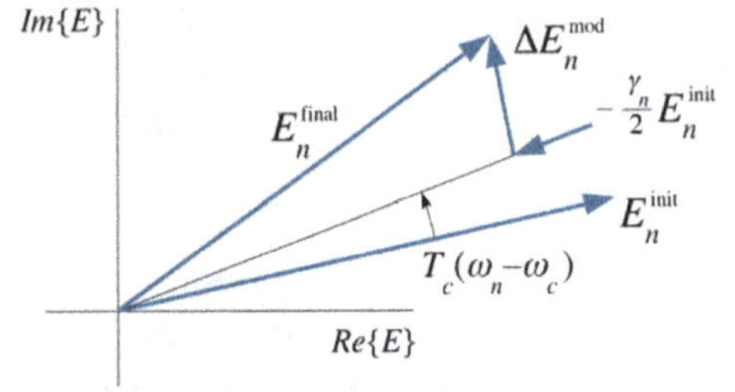

Figure 15.4. Phasor description of active laser mode coupling: $E_n^{\text{final}} = E_n^{\text{init}} e^{iT_c(\omega_n - \omega_c)} - \frac{\gamma_n}{2} E_n^{\text{init}} + \Delta E_n^{\text{mod}}$.

To calculate the sideband modulation ΔE_n^{mod} due to the pulsed electron beam appearing on the rhs of (15.30), we will first consider the effect of a single electron bunch and then consider the effect of a train of electron bunches.

Single electron bunch. Start with the optical pulse evolution equation, (15.9),

$$a(\tilde{z}, \tau) = a_0(\tilde{z}) + \frac{i}{2} \int_0^\tau dp\, j(\tilde{z} + sp) \int_0^p dq \cdot (p - q) \cdot a\big(\tilde{z} + s(p - q), q\big) e^{-i\nu_0(p-q)}, \tag{15.31}$$

and express the microtemporal position $\tilde{z}$ in terms of the time $t = -\tilde{z}/c$. Note that t is not related to τ; the latter represents position along the undulator (~meters), while both $\tilde{z}$ and t represent microtemporal position within the pulse (~microns or ~picoseconds). Graphically, this change of variable simply flips the pulse horizontally, so that the leading edge appears on the left when $a(t, \tau)$ is plotted against t.

In terms of the independent variables (t, τ), we rewrite (15.31), with an implicit change in notation for the functions $\tilde{j} \to j$ and $\tilde{a} \to a$, as

$$a(t, \tau) = a(t, 0) + \frac{i}{2} \int_0^\tau dp\, j\left(t - \frac{sp}{c}\right) \int_0^p dq \cdot (p - q) \cdot a\left(t - \frac{s(p - q)}{c}, q\right) e^{-i\nu_0(p-q)}. \tag{15.32}$$

This equation is again valid for small and large gains and for any degree of optical slippage. For the small-gain amplification of a weak CW optical wave due to a short electron bunch (in which the small-gain assumption is consistent with a time perturbative analysis), we identify the small-gain regime by setting

$$a\left(t - \frac{s(p - q)}{c}, q\right) \to a\left(t - \frac{s(p - q)}{c}, 0\right). \tag{15.33}$$

For a CW input field $a(t, 0) \equiv a_0$ we may also eliminate any reference to the microtemporal position t in the optical field appearing in the integrand.

Extracting the optical field a_0 and integrating to $\tau = 1$ (i.e. to the end of the undulator) we get

$$\Delta a(t) \equiv a(t, 1) - a(t, 0) \tag{15.34}$$

$$= a_0 \frac{\mathrm{i}}{2} \int_0^1 \mathrm{d}p\, j\left(t - \frac{sp}{c}\right) \int_0^p \mathrm{d}q \cdot (p - q) \cdot \mathrm{e}^{-\mathrm{i}\nu_0(p-q)} \tag{15.35}$$

$$= a_0 \frac{\mathrm{i}c}{2s} \int_0^{s/c} \mathrm{d}t'\, j(t - t') \frac{1}{\nu_0^2} \left[\left(\frac{\mathrm{i}\nu_0 ct'}{s} + 1\right) \mathrm{e}^{-\mathrm{i}\nu_0 ct'/s} - 1\right] \tag{15.36}$$

$$\text{or} \quad \Delta a(t) = a_0 \int_{-\infty}^{\infty} \mathrm{d}t'\, j(t - t') g_{\nu_0}(t'), \tag{15.37}$$

where we substituted $t' = sp/c$ in the third line. The time dependence of the dimensionless current density j, (7.28), is given by

$$j(t) = \frac{8\pi^2 e^2 N_w^3 \lambda_w^2 \hat{K}_f^2}{\gamma^3 mc^2} n_e(t), \tag{15.38}$$

and the 'gain function' $g_{\nu_0}(t)$ is defined as

$$g_{\nu 0}(t) = \frac{\mathrm{i}c}{2s\nu_0^2} \left[\left(\frac{\mathrm{i}\nu_0 ct}{s} + 1\right) \mathrm{e}^{-\mathrm{i}\nu_0 ct/s} - 1\right] \quad \text{for } 0 < t < \frac{s}{c}, \tag{15.39}$$

$$= 0 \quad \text{otherwise.} \tag{15.40}$$

For a total charge of eN_e in the electron bunch, the electron density $n_e(t)$ can be normalized as

$$n_e(t) \equiv \frac{N_e}{c\, A_{\mathrm{opt}}} f_e(t); \quad \int_{-\infty}^{\infty} f_e(t)\mathrm{d}t = 1, \tag{15.41}$$

where A_{opt} is the optical mode area. Therefore, the effect of a single electron bunch on a CW optical field is

$$\frac{\Delta a(t)}{a_0} = \mathcal{K} \int_{-\infty}^{\infty} \mathrm{d}t'\, f_e(t - t') g_{\nu_0}(t') \equiv \mathcal{K} f_e(t) * g_{\nu_0}(t) \tag{15.42}$$

where * denotes convolution, and

$$\mathcal{K} = \frac{8\pi^2 e^2 N_w^3 \lambda_w^2 \hat{K}_f^2}{\gamma^3 mc^2} \cdot \frac{N_e}{c\, A_{\mathrm{opt}}}. \tag{15.43}$$

The amplified field of a CW input optical wave is proportional to the convolution of the electron pulse shape $f_e(t)$ with the 'gain function' $g_{\nu_0}(t)$ … a most important result.

Train of electron bunches. We now consider a CW input field of frequency $\omega_k = 2\pi f_k$, and calculate the sideband modulation after a single pass through the undulator due to a train of electron bunches of rep rate f_{RF} separated in time by $T_e = 1/f_{\text{RF}}$.

For the input waveform and input spectrum, we write

$$E_{\text{in}}(t) = E_k \exp(-\text{i}2\pi f_k t) \tag{15.44}$$

$$\tilde{E}_{\text{in}}(f) = E_k\, \delta(f - f_k). \tag{15.45}$$

Note that the sign of ω in our fundamental choice of phase $\text{e}^{\text{i}(kz-\omega t+\phi)}$ implies the temporal Fourier transform pair

$$h(t) = \int_{-\infty}^{\infty} \tilde{h}(f)\text{e}^{-\text{i}2\pi ft}\,\text{d}f \tag{15.46}$$

$$\tilde{h}(f) = \int_{-\infty}^{\infty} h(t)\text{e}^{+\text{i}2\pi ft}\,\text{d}t. \tag{15.47}$$

Now, we had just derived the result that the amplified field of a CW input optical wave is proportional to the convolution of the electron pulse shape $f_e(t)$ with the 'gain function' $g_{\nu_k}(t)$, where the initial phase velocity ν_k in this case corresponds to the input optical frequency ω_k. Thus, if instead of a single electron bunch we have a train of electron bunches separated in time by T_e, we obtain for the amplified output field

$$E_{\text{out}}(t) = E_k \exp(-\text{i}2\pi f_k t)\cdot\left\{1 + \left[\mathcal{K} f_e(t) * g_{\nu_k}(t)\right] * \sum_n \delta(t - nT_e)\right\}. \tag{15.48}$$

Note that only the complex envelope of the amplified field, not the carrier, is modulated by the convolution with the comb function $\sum_n \delta(t - nT_e)$. By successive application of the Fourier convolution theorem and the sifting property of the delta function, we calculate the output optical spectrum to be

$$\tilde{E}_{\text{out}}(f) = E_k\delta(f - f_k) + \mathcal{K} E_k\delta(f - f_k) * \left[\left(\tilde{f}_e(f)\cdot\tilde{g}(f;\nu_k)\right)\cdot\frac{1}{T_e}\sum_n \delta\left(f - \frac{n}{T_e}\right)\right] \tag{15.49}$$

$$= E_k\delta(f - f_k) + \frac{\mathcal{K}}{T_e} E_k\left[\left(\tilde{f}_e(f - f_k)\cdot\tilde{g}(f - f_k;\nu_k)\right)\cdot\sum_n \delta\left(f - f_k - \frac{n}{T_e}\right)\right] \tag{15.50}$$

$$= E_k\delta(f - f_k) + \frac{\mathcal{K}}{T_e}\sum_n E_k\cdot\tilde{f}_e(f_n - f_k)\cdot\tilde{g}(f_n - f_k;\nu_k)\cdot\delta(f - f_n), \tag{15.51}$$

where we define $f_n \equiv f_k + \frac{n}{Te}$. The terms in the summation with $n \neq 0$ represent the sideband components $\delta(f - f_n)$ induced on the input mode E_k by the pulsed electron beam. If many such modes E_k are present, as in the axial mode laser spectrum of the optical resonator, then the total contribution to the mode amplitude E_n for a given f_n is simply the sum over k (including $f_k = f_n$) of all other sideband components coincident with that mode E_n.

The modulation term ΔE_n^{mod} in the mode evolution equation for the amplitude of the nth mode, (15.30), is then

$$\Delta E_n^{\text{mod}} = \frac{\mathcal{K}}{T_e}\sum_k E_k \cdot \tilde{f}_e(f_n - f_k) \cdot \tilde{g}(f_n - f_k\,;\, \nu_k). \tag{15.52}$$

The FEL coupled-mode matrix $\mathcal{M}$. Explicit calculations yield

$$\tilde{f}_e(f_n - f_k) = \int_{-\infty}^{\infty} \mathrm{d}t\, f_e(t) \exp\left[+\mathrm{i}2\pi(f_n - f_k)t\right] \tag{15.53}$$

$$= \int_{-\infty}^{\infty} \mathrm{d}\tilde{z}\, \hat{f}_e(\tilde{z}) \exp\left[-\mathrm{i}(k_n - k_k)\tilde{z}\right] \tag{15.54}$$

where each of $f_e(t)$ and $\hat{f}_e(\tilde{z})$ is normalized, and

$$\tilde{g}(f_n - f_k; \nu_k) = \int_{-\infty}^{\infty} \mathrm{d}t\, g_{\nu_k}(t) \exp\left[+\mathrm{i}2\pi(f_n - f_k)t\right] \tag{15.55}$$

$$= \int_0^{s/c} \frac{\mathrm{i}c}{2\,s\nu_k^2}\left[\left(\frac{\mathrm{i}\nu_k ct}{s} + 1\right)\mathrm{e}^{-\mathrm{i}\nu_k ct/s} - 1\right]\mathrm{e}^{\mathrm{i}(\omega_n - \omega_k)t}\mathrm{d}t \tag{15.56}$$

$$= \frac{\mathrm{i}}{2\nu_k^2}\int_0^1 \left[(\mathrm{i}\nu_k\tau + 1)\mathrm{e}^{-\mathrm{i}\nu_k\tau} - 1\right]\mathrm{e}^{\mathrm{i}(\omega_n - \omega_k)s\tau/c}\mathrm{d}\tau \tag{15.57}$$

$$= \frac{\mathrm{i}}{2\nu_k^2}\int_0^1 \left[(\mathrm{i}\nu_k\tau + 1)\mathrm{e}^{-\mathrm{i}\nu_k\tau} - 1\right]\mathrm{e}^{-\mathrm{i}(\nu_n - \nu_k)\tau}\mathrm{d}\tau, \tag{15.58}$$

where we substituted $(\nu_n - \nu_k) = -\frac{s}{c}(\omega_n - \omega_k)$ in the last line from the frequency dependence of the phase velocity, (11.4). Algebra ultimately yields

$$\tilde{g}(f_n - f_k; \nu_k) = -\frac{1}{2}\left\{\left[\frac{\sin\nu_n}{\nu_n\nu_k} + \frac{\cos\nu_n - 1}{\nu_n\nu_k}\left(\frac{1}{\nu_n} + \frac{1}{\nu_k}\right) - \frac{\cos(\nu_n - \nu_k) - 1}{\nu_k^2(\nu_n - \nu_k)}\right]\right.$$
$$\left. + \mathrm{i}\left[\frac{\cos\nu_n}{\nu_n\nu_k} - \frac{\sin\nu_n}{\nu_n\nu_k}\left(\frac{1}{\nu_n} + \frac{1}{\nu_k}\right) + \frac{\sin(\nu_n - \nu_k)}{\nu_k^2(\nu_n - \nu_k)}\right]\right\} \tag{15.59}$$

$$\equiv -\frac{1}{2}\left[C(\nu_n, \nu_k) + \mathrm{i}S(\nu_n, \nu_k)\right], \tag{15.60}$$

where C, $\mathcal{S}$ are defined in Dattoli and Renieri (1981). Note that this function, and the integral for $\tilde{f}_e(f_n - f_k)$ in (15.54), are complex conjugates of the corresponding functions defined by Dattoli and Renieri (1981), because those authors chose a phase $\exp i(\omega t - kz + \phi)$, which differs from our choice $\exp i(kz - \omega t + \phi)$. The function $C + i\mathcal{S}$ can also compactly be written

$$\mathcal{E}(\nu_n, \nu_k) \equiv C(\nu_n, \nu_k) + i\mathcal{S}(\nu_n, \nu_k) = e^{-i\nu_n}\left[\frac{1}{\nu_n - \nu_k}\left(\frac{1 - e^{i\nu_k}}{\nu_k^2} - \frac{1 - e^{i\nu_n}}{\nu_n^2}\right) + \frac{i}{\nu_n \nu_k}\right]. \tag{15.61}$$

The complete coupled-mode equation, (15.30), can now be written

$$\Delta E_n = \left[-\frac{\gamma_n}{2} + iT_c(\omega_n - \omega_c)\right]E_n - \frac{\mathcal{K}}{2T_e}\sum_k E_k \cdot \tilde{f}_e(f_n - f_k) \cdot \mathcal{E}(\nu_n, \nu_k). \tag{15.62}$$

If we write this equation in the equivalent matrix form describing the evolution of the nth laser mode from pass p to pass $p + 1$ in the resonator,

$$E_n^{(p+1)} = \sum_k \mathcal{M}_{nk} E_k^{(p)}, \tag{15.63}$$

then the elements of the coupled-mode matrix $\mathcal{M}$ are

$$\mathcal{M}_{nk} = \left[1 - \frac{\gamma_k}{2} + iT_c(\omega_k - \omega_c)\right]\delta_{nk} - \frac{\mathcal{K}}{2T_e} \cdot \tilde{f}_e(f_n - f_k) \cdot \mathcal{E}(\nu_n, \nu_k). \tag{15.64}$$

The coupled-mode equation (15.62) for the complex amplitude of the nth mode can also be written in terms of energy and phase, $W_n \equiv |E_n|^2$ and $\varphi_n \equiv \arg\{E_n\}$, as

$$\Delta W_n = -\gamma_n W_n - \frac{\mathcal{K}}{T_e}\sum_k \sqrt{W_n W_k} \cdot \left[\mathcal{B}_{n,k}^C \cos(\varphi_n - \varphi_k) + \mathcal{B}_{n,k}^S \sin(\varphi_n - \varphi_k)\right] \tag{15.65}$$

$$\Delta\varphi_n = T_c(\omega_n - \omega_c) - \frac{\mathcal{K}}{2T_e}\sum_k \sqrt{\frac{W_k}{W_n}} \cdot \left[\mathcal{B}_{n,k}^S \cos(\varphi_n - \varphi_k) - \mathcal{B}_{n,k}^C \sin(\varphi_n - \varphi_k)\right], \tag{15.66}$$

where we defined $\mathcal{B}_{n,k} \equiv \mathcal{B}_{n,k}^C + i\mathcal{B}_{n,k}^S \equiv \tilde{f}_e(f_n - f_k) \cdot \mathcal{E}(\nu_n, \nu_k)$. The functions $\mathcal{B}_{n,k}^{C,S}$ are identical to those defined in Dattoli and Renieri (1981) for a monoenergetic electron beam of zero emittance. Therefore, the phase-energy form of the coupled-mode equations as reported by those authors can be recovered by setting $\varphi_{n,k} \to -\varphi_{n,k}$ and $\Delta\varphi_{n,k} \to -\Delta\varphi_{n,k}$, as required by the opposite choice of phase.

Finally, we note that the term $T_c(\omega_n - \omega_c)$ can be written in terms of the temporal detuning between the electron and optical pulses. Define the *modulation frequency* $\omega_m \equiv 2\pi/T_e$ and the *passive axial mode spacing* $\omega_{ax} \equiv 2\pi/T_c$. The oscillating longitudinal modes ω_n and the corresponding passive normal modes ω_c can then be labeled in terms of the mode number n as

$$\omega_n = \omega_0 + n\omega_m; \qquad \omega_c = \omega_0 + n\omega_{ax}, \tag{15.67}$$

where ω_0 is a reference frequency and n is necessarily the same for each set. The detuning term $T_c(\omega_n - \omega_c)$ can then be written

$$T_c(\omega_n - \omega_c) = T_c\left[(\omega_0 + n\omega_m) - (\omega_0 + n\omega_{ax})\right] \tag{15.68}$$

$$= T_c\, n(\omega_m - \omega_{ax}) \tag{15.69}$$

$$= T_c\, n\left(\frac{2\pi}{T_e} - \frac{2\pi}{T_c}\right) \tag{15.70}$$

$$= \frac{2\pi}{T_e} n(T_c - T_e) \tag{15.71}$$

$$= n\omega_m(T_c - T_e) \to \omega_n \delta T_{cav}, \tag{15.72}$$

where $\delta T_{cav} \equiv T_c - T_e$ and we dropped a constant term $-\omega_0(T_c - T_e)$ upon replacing $n\omega_m \to \omega_n$ in the last line; this is permitted since the term $T_c(\omega_n - \omega_c)$ contributes only to the phase φ_n of the mode E_n, so the omission of a term independent of n merely results in a common shift in the reference phase of the modes.

This completes the derivation of the fundamental equations of FEL supermode theory from a mode locking analysis of the RF-linac FEL. It is instructive to demonstrate, as a final exercise, that the axial modes in the coupled-mode equation, (15.62), are actually uncoupled when driven by a CW electron beam and thus evolve independently in that case.

To construct a CW electron beam in terms of properly normalized electron bunches assumed in the coupled-mode theory, we represent the bunches analytically by the normalized distribution function

$$f_e(t) = \frac{1}{T_e} \operatorname{rect}\left[\frac{t}{T_e}\right]. \tag{15.73}$$

This function represents a rectangular electron bunch of length T_e; thus, a train of such bunches separated in time by T_e will indeed comprise a CW beam. The corresponding number of electrons in each bunch is $N_e = (I/e)_{MKS} T_e$, and the coupling constant $\mathcal{K}$ from (15.43) is

$$\mathcal{K} = \frac{8\pi^2 e^2 N_w^3 \lambda_w^2 \hat{K}_f^2}{\gamma^3 mc^2} \cdot \frac{(I/e)_{MKS} T_e}{cA_{opt}}. \tag{15.74}$$

The Fourier transform $\tilde{f}_e(f_n - f_k)$, obtained by inserting $f_e(t)$ from (15.73) into (15.53), is

$$\tilde{f}_e(f_n - f_k) = \text{sinc}\left[T_e(f_n - f_k)\right]. \tag{15.75}$$

The frequency spacing between the (n, k) laser modes is $f_n - f_k \equiv (n-k)\omega_m/2\pi$, where the separation between adjacent modes is $\omega_m/2\pi = 1/T_e$. Therefore, $\tilde{f}_e(f_n - f_k) = \text{sinc}\,[n-k]$. But the sinc function equals zero at all non-zero integral values of its argument. Consequently, the only surviving term in the summation in the coupled-mode equation, (15.62), is $k = n$, and the modes are indeed decoupled, as we wished to show.

Furthermore, for the self-term $k = n$ in the summation, we have $\tilde{f}_e(f_n - f_n) = 1$, and

$$\mathcal{E}(\nu_n, \nu_n) = C(\nu_n, \nu_n) + \mathrm{i}S(\nu_n, \nu_n) \tag{15.76}$$

$$= -\left[\frac{2 - 2\cos\nu_n - \nu_n \sin\nu_n}{\nu_n^3} + \mathrm{i}\,\frac{2\sin\nu_n - \nu_n - \nu_n\cos\nu_n}{\nu_n^3}\right]. \tag{15.77}$$

Upon substitution of these expressions into (15.62), together with the expression for $\mathcal{K}$ from (15.74), we find that the single-pass complex gain of the amplitude E_n (omitting the cavity loss and detuning terms) is

$$g_n \equiv \frac{\Delta E_n}{E_n} = -\frac{8\pi^2 e^2 N_w^3 \lambda_w^2 \hat{K}_f^2}{\gamma^3 mc^2} \cdot \frac{(I/e)_{\text{MKS}}}{c\,A_{\text{opt}}} \cdot \frac{\mathcal{E}(\nu_n, \nu_n)}{2} \tag{15.78}$$

$$= +j\left[\frac{2 - 2\cos\nu_n - \nu_n\sin\nu_n}{2\nu_n^3} + \mathrm{i}\,\frac{2\sin\nu_n - \nu_n - \nu_n\cos\nu_n}{2\nu_n^3}\right], \tag{15.79}$$

where the expression for j for a filamentary electron beam is given by (7.28) with $n_e = (I/e)_{\text{MKS}}/cA_{\text{opt}}$. This is precisely the result we calculated from the FEL weak-field equation in section 14.2.

We end our journey on an historical note. Two regimes of FEL operation were recognized early on. The first was the Raman regime, defined by high current, low energy electron beams in which the electron motion is influenced both by the optical field and by collective space charge effects. The second was the Compton regime, defined by low current, high energy electron beams in which space charge forces are generally negligible. The Mark III FEL, for example, is a Compton regime device, as are most FELs operating today. The Compton regime is also known as the 'single particle' regime, since the phase space evolution of individual electrons can be calculated independently of one another in the presence of the optical field.

The classical theory of the Compton FEL was developed on the basis of two separate approaches contemporaneously and independently in the late 1970s and early 1980s. The first was based on the equations of Maxwell and Lorentz in the time domain (Colson 1981) leading to the coupled FEL equations of motion, and the second was based on a multimode Hamiltonian analysis in the frequency domain (Dattoli and Renieri 1981) leading to the supermode theory.

The observation that the supermode theory, when applied (as we did above) to a CW electron beam, yields the identical complex gain obtained from the coupled FEL equations of motion suggests that the former theory, when transformed into the time domain, is physically equivalent to the latter. This equivalence is further revealed by the pulse evolution analysis of (15.13)–(15.19) and the cavity detuning results of (15.27) and (15.20). It is thus satisfying to see that the connection can be 'made the other way'—that it is possible to show that the coupled-mode representation of the supermode theory in the frequency domain, (15.62; 15.65; 15.66) can be derived from the coupled FEL equations of motion in the time domain, (15.2; 15.3; 15.9) through a straightforward application of mode locked laser theory.

References

Colson W B 1981 The nonlinear wave equation for higher harmonics in free-electron lasers *IEEE J. Quantum Electron.* **QE-17** 1417–27

Dattoli G and Renieri A 1981 The free-electron laser single-particle multimode classical theory *Nuovo Cimento* B **61** 153–80

Dattoli G, Giannessi L, Renieri A and Torre A 1993 Theory of Compton free electron lasers *Progress in Optics* vol 31, ed E Wolf (Amsterdam: North-Holland)

Siegman A E 1986 *Lasers* (Mill Valley CA: University Science Books)

Szarmes E B 1992 High resolution free-electron laser spectroscopy *PhD thesis* Stanford University Palo Alto, CA

Szarmes E B and Madey J M J 1993 The Michelson resonator free-electron laser. Part II: Supermode structure and mirror detuning effects *IEEE J. Quantum Electron.* **29** 465–78

www.ingramcontent.com/pod-product-compliance
Lightning Source LLC
Chambersburg PA
CBHW082035230326
41598CB00081B/6522